A History in the Making

Other McGraw-Hill Aviation Titles

U.S. Civil Aircraft Series, Vols. 1–9
Joseph P. Juptner

American Aviation: An Illustrated History, Second Edition
Joe Christy and Leroy Cook

The Aviation Fact Book
Daryl E. Murphy

Airplane Maintenance & Repair:
A Manual for Owners, Builders, Technicians, and Pilots
Douglas S. Carmody

Aircraft Partnership
Geza Szurovy

The Cessna 150 and 152, Second Edition
Bill Clarke

The Cessna 172
Bill Clarke

The Piper Indians
Bill Clarke

The Piper Classics
Joe Christy

The Complete Guide to Single Engine Cessnas, Second Edition
Joe Christy and Brian J. Dooley

A History in the Making

80 Turbulent Years in the American General Aviation Industry

Donald M. Pattillo

McGraw-Hill

New York San Francisco Washington, D.C. Auckland Bogotá
Caracas Lisbon London Madrid Mexico City Milan
Montreal New Delhi San Juan Singapore
Sydney Tokyo Toronto

Library of Congress Cataloging-in-Publication Data

Pattillo, Donald M.
A history in the making : 80 turbulent years in the American general aviation industry / Donald M. Pattillo.
p. cm.
Includes bibliographical references and index.
ISBN 0-07-049448-7 (alk. paper)
1. Aeronautics—United States—History. 2. Aircraft industry—United States—History. I. Title.
TL521.P38 1998
338.4'762913'00973—dc21 98-4213
CIP

McGraw-Hill

A Division of The ***McGraw-Hill*** *Companies*

1 2 3 4 5 6 7 8 9 0 DOC/DOC 9 0 3 2 1 0 9 8

ISBN 0-07-049448-7

The sponsoring editor for this book was Shelley Carr, the editing supervisor was Jane Palmieri, and the production supervisor was Pamela Pelton. It was set in in the GEN1-AV dseign in Garamond by Dennis Smith of McGraw-Hill's desktop publishing department, Hightstown, N.J.

Printed and bound by R. R. Donnelley & Sons Company.

To Sharon and Duncan

Contents

List of Tables

Preface

Aviation history is a highly developed field. Numerous works cover specific periods such as the Golden Age of 1919–1939; famous aircraft and major aircraft classes; histories of such major firms as Boeing, Lockheed, and McDonnell Douglas; and biographies of leading personalities such as the Wright brothers, Lindbergh, and Earhart. There are also histories of major aeronautical developments such as the jet engine and the helicopter, of record-breaking flights, and of other aeronautical feats. In addition, there are scholarly studies of the societal and economic impact of aviation and of its technical development. No aspect of aviation history has been ignored by scholars or aviation professionals.

A conspicuous gap in the literature, however, is a comprehensive, balanced historical survey of the general aviation manufacturing sector of what is now the aerospace industry. There are, of course, numerous books on private flying and published histories of such companies as Piper, Beech, Cessna, Taylorcraft, and Luscombe, but there has been no overall survey history of general aviation manufacture. This gap stands in marked contrast to the extensive literature on the large military and commercial aircraft sectors. Yet the general aviation sector predates both, extending to the beginnings of powered flight, and has been particularly characterized by a spirit of entrepreneurship. General aviation also differs from the much larger military, commercial, and space sectors in that it more closely approaches a consumer products industry, marketing a range of aircraft for individual customers rather than scores or hundreds of a single type to a large airline or military service. General aviation dominates in such important measures as numbers of active aircraft, airfields, pilots, and flying hours, yet it is the least understood sector of the industry. A major reason is a lack of historical perspective.

This work bridges that gap in both aviation and business history. Yet defining which firms should be included in a history of the general aviation sector was not always clear. Leading companies like Beech and Cessna have important military operations, and indeed would have been unlikely to survive without that strong military base. Conversely, many large military and commercial firms have been involved in general aviation as well. Most general aviation helicopters, in fact, have been developed by primarily military firms. Their activities in civil helicopters have been allotted extensive coverage in this work.

An overlap in civil and military design has always existed, as business and personal aircraft have been adapted for military training and utility use, while certain military and commercial designs have been adapted as executive transports, firefighting aircraft, and for other civil roles. That overlap is particularly evident in smaller helicopters. Accordingly, certain determinations were necessary as to which firms and operations were included. Further, given the numerous attempts at designing, building, and selling light aircraft, only those firms judged to have been significant, either in production or innovation, have been included. Thus many may find an obscure or short-lived firm of their knowledge missing.

Maintaining a manageable focus also presented a challenge. In more recent years, the homebuilt or kit industry segment has gained in significance, largely succeeding the factory-built small trainer aircraft segment. Several firms now offer innovative, successful kit designs. Accordingly, that segment is given substantive treatment. But ultralight/microlight sport aircraft, sailplanes, racing aircraft, and experimental prototypes generally are excluded. While undoubtedly significant to many, these activities are very narrow segments of general aviation, which is primarily oriented toward business flying and other revenue-generating activities. Moreover, microlights do not require FAA certification or pilot licensing. Also generally excluded are modification and conversion firms, as they do not involve original design and production.

The general aviation industry has experienced a turbulent history, with a very high failure rate. But general aviation is indeed a national resource with a significant impact on the economy. It remains an underappreciated aspect of the national aviation heritage. It is the author's hope that this work will be a step toward fuller understanding.

Donald M. Pattillo

Acknowledgments

This history of general aviation aircraft manufacturing began as a component of a larger and more comprehensive history of the American aircraft industry. As work progressed, however, it became a concern to many with whom I had contact that the eventual work would be too lengthy and unwieldy to be widely marketable. Accordingly, I separated the general aviation sector from the overall work and undertook its development as a separate history. While reluctant initially, I soon became completely committed to the project, spurred by the knowledge that such a work had never been accomplished previously. In addition, it became apparent that a separate treatment of general aviation manufacturing was logical in that the structure, size, and market for general aviation aircraft were entirely different from those of the large military and commercial aircraft manufacturers.

As I began to extend and expand the coverage of general aviation history, I sought and received help from many sources and many individuals. Without that assistance the work would not have been completed, or certainly would not have been of the quality it attained. In no particular order, I feel indebted to Pat Reilly of the New Jersey Aviation Museum; to Dr. Roger Bilstein of the University of Houston, Clear Lake City; to Dr. William F. Trimble of Auburn University; to Dawne Dewey of the Special Collections and Archives of the Library of Wright State University; to Ronald E. Green of the Aerospace Policy and Analysis Division, International Trade Administration, U.S. Department of Commerce; to librarian Mary Jane Townsend of the Kansas Aviation Museum; and to Mike Kelly and Mary Nelson of Special Collections, the Ablah Library, Wichita State University.

I have received specific and continuing assistance from Shelly Snyder of the General Aviation Manufacturers Association. I have also benefited from a discussion with Ed Phillips of *Aviation Week & Space Technology*. Ned Preston, historian of the Federal Aviation Administration, provided generous assistance. Mrs. B. D. Maule of Maule Aircraft was very generous with assistance about her company. Several individuals at the Experimental Aircraft Association provided information in their area of expertise.

I owe a special thanks to my wife Sharon, who as a professional librarian helped me with numerous data sources, particularly computer data searches, with which I remain rather clumsy. She also has been very supportive in my relentless demands on what should have been her leisure time in that regard.

Finally, but perhaps most important, I am deeply indebted to Shelley Carr, acquisitions editor of the Aviation Program of McGraw-Hill. Without her interest, active support, encouragement, and advice, this book would not have become what it is. Her assistant Gwen Myers also was extremely helpful. Further thanks are due to Jane Palmieri, senior editing supervisor, McGraw-Hill, for the professional and expeditious manner in which she moved the manuscript through to publication.

I have endeavored to achieve factual accuracy and to present sound analysis and interpretation of conditions and events. In that effort I have of course relied, in varying extents, on the work and views of others. But as the author, the responsibility for any factual errors, omissions, or flaws in analysis or inference is mine and mine alone.

A History in the Making

1

Origins of Personal Aviation

In the early years of powered flight, as aircraft first began to be produced in series, almost all aviation activity was encompassed within what would later be defined as general aviation. There were no well-defined military roles for aircraft and no scheduled commercial services. Thus the market for aircraft was largely limited to sport flying and training by and for a wealthy few. But the First World War saw military applications developed and implemented, and the military sector dominated aviation. Then with the return of peace, many saw the potential for a large civil sportplane market, building on major technical progress and on the increased awareness of aviation brought about by the war. Other aircraft categories and roles to become included in general aviation were not then foreseen and thus had not evolved. But as this opening chapter illustrates, many developments important to the future of aviation, and to personal aviation specifically, occurred during the 1920s.

The Light Aircraft Market

Efforts to fill the anticipated peacetime market for sport flying and new sport aircraft were made on both coasts. The inventor Lawrence Sperry of New York and the Loughead (later Lockheed) brothers of California were among the first to develop small sportplanes. The Loughead S-1 monoplane of 1919, designed by John Northrop, had many advanced features, as did the Sperry M-1A Messenger, a single-seat biplane. Both were worthy designs but did not sell given the glut of surplus wartime models on the small civil market, although the Army procured 20 Messengers in 1921 as communications aircraft. Sperry, tragically, died in a crash on December 13, 1923, at age 31, also ending his company's prospects. The Loughead firm suffered bankruptcy in 1921.

While a market for new private aircraft did not develop in the early 1920s, wartime aeronautical progress still conveyed major benefits to private aviation: the number of pilots had increased rapidly, and there had been enormous advances in aircraft production techniques. The open-cockpit biplane constructed of wood and fabric still dominated, but the abundance of surplus wartime trainers, especially the ubiquitous Curtiss JN-4D Jenny, enabled many to learn to fly relatively cheaply. Interest in private flying was further stimulated in the early 1920s by the barnstorming era and by the popular military-sponsored air races. Barnstormers, or "flying gypsies," traveled from place to place conducting flying exhibitions in the Jenny and comparable aircraft. But barnstorming, while undoubtedly promoting interest in aviation, declined after the mid-1920s. Further, flying was still publicly regarded, not inaccurately, as dangerous. Yet the groundwork had been laid for a permanent private aviation sector.

Development of scheduled airline service began roughly concurrently with growth of the private aviation sector. But development of efficient, safe, and comfortable passenger-carrying aircraft was painfully slow. Scheduled passenger services in fact lagged air mail, the first commercial growth area, by several years.

All commercial and private aviation, or civil aviation, began to be encompassed within a somewhat nebulous aerial service sector. While lacking precise definition, aerial service involved much more than sport or exhibition flying. Among the first to see a broader potential for light aircraft was Sherman Fairchild of New York, who pioneered not only aerial photography but also development of specialized photographic planes. The State Experimental Station of Ohio first tested airplanes for agricultural use in August 1921. Lieutenant J. A. Macready of McCook Field, a record-breaking Air Service pilot, conducted the experiments.[1] A new firm, Huff Daland Airplanes of Ogdensburg, New York, first adapted an airplane for aerial cropdusting in 1924. The company thereupon founded Huff Daland Dusters in Macon, Georgia, as the first aerial cropdusting service, using its own designs. The present-day Delta Air Lines sprang from that service. Huff Daland also experimented with, but did not produce, personal aircraft.

Despite innovation and development, the market for new airplanes remained depressed. Public concerns about flying safety further inhibited market expansion. The government had done little to develop civil aviation policy, and there was a lack of licensing, training standards, and airways development. Then-active aero clubs, led by the New York Aero Club, developed and administered their own standards and regulations. Further, the light aircraft manufacturing sector was too small to exert political influence. Both the Manufacturers Aircraft Association (MAA) and the Aeronautical Chamber of Commerce (ACC) focused chiefly on military aviation and commercial, primarily air mail, operations. The National Advisory Committee for Aeronautics (NACA), the government agency conducting research and experimentation in the field, did little specifically to advance personal aviation or light aircraft development.

Another deterrent to those potentially attracted to the field was that light aircraft design generally overlapped technically with military aircraft. Military-oriented firms could and did serve both the military and civil sectors. Curtiss, for example, while a leading military firm, was among those active in private aircraft. Accordingly, a firm wishing to specialize in light civil aircraft could find itself at a disadvantage in resources.

In the midst of dismal financial and market prospects faced by all aircraft manufacturers in the mid-1920s, the President's Aircraft Board, popularly known as the Morrow Board and established in reaction to Billy Mitchell's charges of neglect of air power, carried an impact. The Morrow Board report at the end of 1925, while emphasizing military needs, spurred passage of the Air Commerce Act of 1926. Signed into law by President Coolidge and implemented by Secretary of Commerce Herbert Hoover, the Act benefited all civil aeronautics, including personal flying and other aerial service activities. Aircraft safety and airfield development advanced, and the first federal pilot's licenses were issued in 1927. William P. MacCracken, head of the MIT aeronautical engineering program, was named first chief of the new Aeronautics Branch, and was issued license No. 1 on April 6, 1927. The Act also enabled reliable statistics to be compiled on flying and pilots for the first time.

The aerial service sector concurrently gained more precise definition, to include industrial uses such as crop dusting, aerial photography and survey, forest fire patrol and wildlife management, as well as air taxi work and flying instruction.[2] Air mail service continued to grow, as did sport or recreational flying, as the number of pilots increased. Skywriting, oil and timber survey, and emergency medical transport roles also appeared. Another aerial service function which emerged was that of news coverage, as aircraft could reach the site of a major event or disaster faster than other means of transport. Scheduled passenger service, while embryonic, was also considered within aerial service, as the personal and commercial sectors did not become largely separate until the early 1930s.

Fixed-base operators, primarily involved with training and maintenance, were established from 1920. Such operations facilitated growth of the sector, especially cross-country flying. Partially reflecting the rather primitive technology of the day, however, was that personal aircraft still used automobile gasoline. Specially refined aviation fuel had not yet appeared.

One private aviation function, little discussed at the time and for years afterward, was the transportation of illegal liquor during Prohibition. The pilot Ben O. Howard, later to gain fame in air racing and as a designer, admitted in later years to aerial bootlegging during the 1920s.[3]

Another civil aircraft role beginning in the 1920s, which was to attain near-legendary status in decades to come, was that of bush flying. The term *bush,* originally referring to the scrublands of South Africa, was extended to include all remote or wilderness regions. Originating in northern Canada, Alaska, and remote regions of the Pacific Northwest, bush flying eventually extended to Australia, New Guinea, Mexico, and South America. Roles involved not only transportation of hunters, fishermen, and prospectors but aerial mapping, transportation of vital food and medical supplies, and emergency evacuations from sites unreachable by land or water. Float- and ski-landing gear largely developed from bush flying.

Business flying, however, still was very rare: only the largest corporations and a few wealthy businessmen used personal aircraft. But business or executive aviation began to expand from the mid-1920s, and the Loening Air Yacht, a comfortable enclosed-cabin design powered by a pusher Liberty engine, was a popular business amphibian. The Douglas Dolphin of 1929, with six to eight seats and twin engines, also was a successful business amphibian, and was ordered by the Army, Navy, and Coast Guard as well. President Roosevelt used a Dolphin during the 1930s on official travel. Keystone Aircraft (formerly Huff Daland), after acquiring Loening in 1928, also produced a civil Air Yacht, although unrelated to the Loening design.

Personal Aircraft Entrants

Despite the virtually nonexistent market for new light aircraft in the early 1920s, production ventures, based more on future optimism than response to current demand, were formed. Among the earliest firms specializing in light aircraft was that of the barnstormer George E. "Buck" Weaver, who had worked for the early East Coast aircraft firms L-W-F and Aeromarine, and his brother-in-law Charles W. Meyers. The partners first located in Lorain, Ohio, late in 1919 to test Weaver's design. The partnership was dissolved early in 1920, but a successor, Weaver Aircraft Company, emerged in 1921 at Medina, Ohio. Weaver brought in two acquaintances, Clayton J. Brukner, with experience at Curtiss and other East Coast aircraft firms, and his friend Elwood James "Sam" Junkin, as investors. Both contributed to design and production. After financial struggles Weaver was reorganized as Advance Aircraft Company on February 26, 1923, in Troy, Ohio, with further financial backing from the wealthy Alden Sampson II. Junkin became president. Buck Weaver departed the company and died in July 1924, but Advance Aircraft, trading under the acronym WACO, continued his designs and became one of the most successful producers of open-cockpit biplanes.

Sampson later withdrew his interest; then Junkin, who had married Weaver's widow, died on November 1, 1926. But the company progressed during the late 1920s under Brukner as majority owner, and by avoiding debt was able to survive the depression. In June 1929 it floated a public stock offering as WACO Aircraft Corporation and was numerically the largest producer in the industry. From 1930 all designs were given letter designations, and the product line included cabin monoplanes as well as biplanes.[4]

E. M. (Matty) Laird, a young aviator and businessman from Chicago and a friend of Buck Weaver, located in Wichita, Kansas, in the summer of 1919 to produce personal aircraft. This marked the first manufacturing venture in the city which was to become the light aircraft or general aviation capital. Wichita, enjoying an oil boom at the time, had investors available, and Laird attracted support from local businessmen W. A. (Billy) Burke and Jacob Moellendick of $15,000 each, while he contributed designs, materials, and experience. The group purchased Moellendick's Wichita Aircraft Company, formed only in July, in December 1919. The E. M. Laird Company was formed in May 1920, and Wichita Aircraft Company was dissolved.

The Laird Swallow biplane, designed with the assistance of Buck Weaver, was tested in April 1920 and attracted immediate orders. A new factory was completed in late 1921. Matty Laird was soon joined by his brother Charles, and then by two former military aviators, Lloyd Stearman and Walter Beech. Weaver continued to help Laird during a time when his own venture was floundering, but was fired by Moellendick late in 1921. Laird produced the Swallow until September 1923 when, after conflicts with Jacob Moellendick over expansion plans, he sold his interest and returned to Chicago.[5] Then on January 22, 1924, Moellendick formed Swallow Airplane Manufacturing Company, named after its major product, and continued production. The improved New Swallow, designed by Lloyd Stearman, was also employed on air mail services from 1926.

Swallow suffered a major setback when its entry into the California-Hawaii air race was lost, and it declared bankruptcy on August 12, 1927, ruining Moellendick financially. The firm was reorganized as Swallow Airplane Company on November 15, 1927, after Moellendick sold rights to Lincoln, Nebraska, investors led by Victor Roos.[6] Swallow sales recovered. Charles Laird established concurrently a separate Laird Aircraft Corporation with himself as president. The firm traded as Whipporwill in order to avoid confusion with the new E. M. Laird Company in Chicago, which Matty Laird had organized in 1926.[7] Whipporwill found little market success with its design, however, and succumbed to the depression and bankruptcy on December 31, 1930. Swallow survived as a corporation, but also exited aircraft production during the depression.

Although Moellendick is a largely forgotten figure in aviation and appears to have been rather difficult as a business associate, he probably deserves more credit than any other individual for establishing Wichita as the center of the light aircraft industry. While he grew wealthy in the oil boom, Moellendick suffered from alcoholism as well as business reverses and died penniless in 1940. But city support, topography, climate, and a trained and experienced labor force enabled Wichita to attract and retain numerous private aircraft firms. The later success of Beech and Cessna secured its claim as the aviation capital. Stearman, a manufacturing division of Boeing Airplane after Boeing became independent of United Aircraft in 1934, later led Wichita into military aircraft production as well.

To have an enduring impact on personal aviation was the Travel Air Corporation, formed late in 1924 and incorporated on February 4, 1925. Walter H. Beech, a former Army Air Service pilot and barnstormer, founded the firm with several associates, including Lloyd Stearman, and served as general manager. While with Laird, Beech and Stearman favored steel airframes for the New Swallow, which Moellendick opposed, and both departed early in 1924. Beech then capitalized on numerous aviation contacts made during his wartime service in organizing the new venture.

Businessman Walter P. Innes Jr., owner of Wichita's largest department store, arranged financial backing from Hayden, Stone and Co., and served as president. Innes also suggested the name Travel Air. Clyde V. Cessna, a pioneer pilot, designer, and builder, joined as vice president. Successful in farming at the time, Cessna also was a major investor. Mac Short, an MIT engineering graduate and, like Stearman, a native Kansan, joined, and Charles Yankey, another Wichita native and friend of Beech, was an original investor.[8] Beech, while neither a designer nor engineer, was the driving force in management. Joining the small staff as company secretary and bookkeeper was 21-year-old Olive Ann Mellor. In 1930 she would become Mrs. Walter H. Beech.

The 2000/3000/4000 series of Travel Air open-cockpit biplanes was one of the most successful of the period. They were later joined by the 5000/6000 series of 6-seat cabin monoplanes, which Clyde Cessna helped design. The Model 6000 also was offered with Edo floats, as floatplane use expanded in the 1920s. (Earl D. Osborn patented aircraft floats. He incorporated his Edo firm on Long Island on October 29, 1925, and began production in 1926.)

Walter Beech became president of Travel Air on February 15, 1927, and was firmly in control of the company. By that time Stearman had left to form his own company, and Cessna soon followed. In 1929 Travel Air sold a total of 547 aircraft and was the largest producer in Wichita.[9] The factory had expanded to 116,000 square feet. R. K. Beech, younger brother of Walter, joined the firm in 1928. Beech then sold Travel Air to Wright Aeronautical Corporation, also financed by Hayden, Stone and soon to become Curtiss-Wright, in June 1929, providing stockholders with a major profit.[10] The timing was sheer luck, as the stock market crash soon thereafter would have largely destroyed Travel Air's value.

The entrepreneur J. Don Alexander moved to Englewood, Colorado, a Denver suburb, where he established Alexander Industries, principally for production of motion picture film. He became determined to build an airplane as good as the Laird Swallow, and which would perform well at high elevations. He thereupon formed Alexander Aircraft Corporation in August 1925 as a subsidiary of Alexander Industries. Alexander employed the youthful Al Mooney as a designer. Mooney gave up a chance to study engineering for the opportunity. Mooney designed the successful Eaglerock two-seat biplane and went on to design the advanced Bullet cabin monoplane. Alexander moved production to a new factory in Colorado Springs in April 1928, and was a leading light aircraft manufacturer in the expanding market of the late 1920s. The Bullet won substantial orders, but it experienced stability problems and few were delivered. Alexander also became a victim of the depression and went bankrupt in August 1932. Al Mooney had left earlier, in May 1929, to form his own firm in Wichita. He tested his M-4 design later that year, but the depression ended production efforts in 1931.

Lloyd Stearman, formerly a naval aviator, experienced a varied career in Wichita with Laird, Swallow, and Travel Air. But Stearman, in common with many others in the early industry, aspired to run his own company, and departed Travel Air in October 1926. He founded Stearman Aircraft, Incorporated, in Venice, California, by taking over the West Coast distributor for Travel Air.[11] Mac Short joined him. Stearman's first design, the C-1 three-place biplane, powered by the old OX-5 engine, won favorable attention. He entered production in Venice with the successor C-2 but soon encountered financial difficulties. Attracted again to Wichita with financial assistance from Walter Innes, he established Stearman Aircraft Corporation on September 27, 1927. Mac Short followed, joining him as vice president, and a new factory was built. Production of Stearman's C-3 biplane series built up rapidly in the boom of the late 1920s.[12]

The young investment banker Robert Gross first met Stearman in 1928 while evaluating his firm as a financing client. Given his first airplane ride by Stearman, Gross was motivated to cast his future with aviation and became a financial advisor to Stearman. Gross later built Lockheed into a major firm. On August 15, 1929, Stearman Aircraft joined other firms as a component of the new giant, United Aircraft and Transport Corporation (UATC). The transaction was in the form of a stock swap, however, and with the crash Lloyd Stearman was left with only his salary from UATC.

Alfred V. Verville, a major designer and innovator from the First World War period to the 1930s, enjoyed a richly varied career as a designer and manufacturer. He began with Glenn Curtiss in 1914, then went briefly to the Thomas Brothers firm in 1915. He formed his General Aviation Company in April 1915 but the venture lasted only two years. He worked for Fisher Body during the war period, then for the Air Service Engineering Division helping with pursuit and racing aircraft development. He also collaborated with Lawrence Sperry on the Messenger and other postwar designs. Verville resigned from the Engineering Division in February 1925 and joined Lawrence D. Buhl to form the Buhl-Verville Aircraft Company in Michigan. The company enjoyed some success with the Airster cabin monoplane design, but in 1927 Verville sold his interest to Buhl, who continued development.

Forming a new Verville Aircraft Company, he achieved a notable success with the Air Coach cabin monoplane. In 1930 Verville entered and won the Army trainer competition with his PT-10, adapted from a civil model. But he could not arrange financing to fulfill the order and became another bankruptcy victim in 1931.[13] While never a major manufacturer, Verville's design abilities nonetheless won him a place in industry history.

Edward A. "Eddie" Stinson, who learned to fly with the Wright brothers in 1911, earned renown as an instructor, stunt flyer, and record-breaking pilot. He established a flight school with his flying sisters in December 1915. He formed the Stinson Airplane Company in 1920 in Dayton, but his major business activity would be in Detroit, Michigan. The Aviation Committee of the Detroit Board of Commerce backed Stinson in forming the Stinson Airplane Syndicate, which located in suburban Wayne in 1925. Many of his backers later were involved in the formation of the ill-fated Detroit Aircraft Corporation.

Stinson was provided with $25,000 to develop an inherently stable cabin monoplane. His design, the six-seat S.M. 1 Detroiter, first flew on January 25, 1926. The Detroiter was advanced for its time, with a quiet, fully enclosed heated cabin, engine starter, and wheel brakes. During its career it broke or established numerous flight records. With the success of the Detroiter, Stinson formed the Stinson Aircraft Corporation on May 4, 1926, with public capital of $150,000.[14] Stinson by then also was earning $100,000 a year as a stunt pilot. The company sold 10 Detroiters in 1926 and pursued further developments of the basic design.

Including exports, Stinson delivered 121 airplanes in 1929, and in September of that year the automobile magnate E. L. Cord, increasingly involved in aviation, acquired 60 percent of the stock. With Cord's support, the Stinson 6000 trimotor airliner, originally the Corman 3000,* achieved some success and also underpriced the successful Ford trimotor. Cord employed the trimotor in his airline operations. Stinson developed the S.M. 2 Junior, a smaller version of the Detroiter, in 1929. Stinson's sales held up better than most in the depression, and in 1930 he offered six models, ranging from the four-seat Junior to the trimotor airliner. Successive designs were produced well into the 1940s under later corporate owners Aviation Corporation and Consolidated Vultee. Stinson did not live to see the later success of his company, however, as he died in an air crash at Chicago on January 26, 1932, while on a sales trip. Only 38, he had accumulated some 16,000 flying hours, more than any other person.

* Corman Aircraft, Inc., of Dayton, Ohio, was a venture of E. L. Cord and Lucius P. Manning; hence the name.

Sherman Fairchild determined to develop a specialized airplane for aerial photography. For that purpose, and with the benefit of family wealth, he formed Fairchild Aviation Corporation in 1925 in New York. This was a separate undertaking from his other interests. Fairchild's FC-1 (for Fairchild cabin model) photographic airplane, designed with the assistance of Igor Sikorsky, first flew on June 14, 1926. The FC-1 and successive variants became popular for bush flying as well. Fairchild also acquired the Caminez Engine Company of Ohio in 1925 as the Fairchild Engine Company.

Fairchild Aviation grew to hold two manufacturing arms, Fairchild Airplane Manufacturing Corporation at Farmingdale, Long Island (in the old Sperry factory), and early in 1929 Kreider-Reisner Aircraft Company of Hagerstown, Maryland.[15] Kreider-Reisner had been incorporated on September 5, 1925, by Ammon H. Kreider and Lewis Reisner as the East Coast distributor for Waco but soon began manufacturing their developments of Waco open-cockpit designs. Fairchild purchased a majority of Kreider-Reisner stock on April 1, 1929, but Kreider remained as president and the Hagerstown factory was expanded.[16]

Fairchild enjoyed strong aerial mapping business, and with overall growth consolidated aviation operations under a reorganized Fairchild Aviation Corporation, incorporated in Delaware in 1927, thereby becoming a somewhat vertically integrated holding company in aviation.[17] Acquisition of K-R enabled Fairchild to broaden his product line and to increase overall aircraft production. Kreider was killed in a mid-air collision on April 13, 1929, however, and Reisner departed thereafter, but the K-R designs continued under the Fairchild nameplate. The Kreider-Reisner Challenger C-6 light biplane, modified from the Waco Model 9, became the Fairchild KR-21.

Ultimately to become one of the most successful manufacturers in the field was Clyde V. Cessna. Cessna had first flown in 1911, and he established a small shop in Wichita in 1916 to construct his own design.[18] His Comet monoplane of 1917, the first airplane ever built in Wichita, held promise, but Cessna failed to sell the design to the War Department and subsequently returned to farming for several years.[19] Reentering the industry in 1924 with Travel Air, Cessna eventually differed with Walter Beech, who emphasized biplanes while Cessna was determined to develop fully cantilevered monoplanes. Cessna primarily wished to start his own company, however, and he and Beech remained on good terms. Leaving in April 1927, Cessna sold his Travel Air stock to outside investors and incorporated his own firm on September 8, 1927, in Wichita.[20] Cessna was the major stockholder, but the second largest investor was businessman Victor Roos. Another original stockholder was Clement Keys, president of Curtiss. The firm was briefly named Cessna-Roos, but Roos resigned in November 1927 after taking control of Swallow and becoming general manager, and the firm became Cessna Aircraft Company on December 31, 1927.

Cessna began his A-series of high-wing monoplanes in 1928, producing 50 of the model AW, followed by 117 of the later model AA, excellent business for a new firm. A significant further development was the CW-6 (for third series, Wright power, and six seats), which flew in November 1928, followed by the more refined DC-6 with a Curtiss Challenger engine, which cost in the $11,000 range.

There were numerous short-lived private aircraft ventures during the decade of the twenties, although details of their organization and management frequently are obscure. The Buffalo firm of G. Elias and Brother, Inc., established in 1881, entered aircraft manufacturing in 1919. It produced sporadically a series of biplane military trainers and civil cabin airplanes through the 1920s. It was succeeded by an autonomous Elias Aircraft and Manufacturing Company in 1929, but the firm disappeared in the depression.

The Gates-Day Aircraft Company was chartered on October 17, 1927, in Paterson, New Jersey, to manufacture an updated version of the wartime Standard J-1 trainer for the civil market. Ivan Gates, owner of the Gates Flying Circus and a successful exhibition flyer, was president, and Charles H. Day, a wartime engineer for Standard, was chief engineer. The four-seat Gates-Day D-23 was followed by the GD-23 and GD-24 versions, all open-cockpit biplanes. After policy differences occurred, Gates withdrew in April 1928 and Charles L. Augur came in as president, bringing stronger financial backing. The company name was changed to the New Standard Aircraft Company on December 29, 1928, still marketing developments of the J-1.[21] Over 60 of the improved D-25 model were built, but the Great Crash weakened the company. A few NT-1 naval trainers were delivered in 1930 but could not assure the company a future. Day became president in 1930, but he sold his interest in April 1931. New Standard, along with many others, went bankrupt later in 1931. Day and his wife then won acclaim for their around-the-world trip, which they completed in December 1931 in a New Standard airplane.

The Aerial Service Corporation was founded in Hammondsport, New York, in 1920. It produced a series of light aircraft under the Mercury nameplate and was later renamed Mercury Aircraft. Small-scale production extended into the 1930s. The Cox-Klemin Aircraft Corporation was founded at College Point, Long Island, in 1921. A collaboration of Charles Cox and Professor Alexander Klemin of New York University, an early graduate of Dr. Jerome Hunsaker's aeronautics program at MIT, the firm produced biplane Army trainers and scout floatplanes for the Navy. It also developed an experimental two-place amphibian design. Cox-Klemin took over the wartime Ordnance/Baldwin plant on Long Island in 1924, but Klemin soon resigned. Unable to sustain production, the business faded in 1925.

One short-lived private aviation venture, which nonetheless attained near-legendary status, was Heath Aircraft Company of Chicago, founded in 1926 by Edward A. Heath. Beginning just when the supply of cheap war surplus trainers was drying up, Heath believed that he could offer a reliable aircraft with a cost low enough to sell in the limited market. The Heath Parasol, a small single-seater somewhat similar to later ultralight aircraft, and powered by a 27 hp motorcycle engine, first flew in 1926. It represented one of the earliest entries into what eventually became known as the homebuilt field.* The Parasol was produced both fully assembled and in kit form from 1927 to 1931. Complete aircraft at a fly-away cost of $975 and kits, with a base price of $199, enjoyed brisk sales. Ed Heath did not live to see the later growth of the field, however, as he was killed in a crash in February 1931, and his company succumbed to the depression.

# # # # #

* Aircraft plans for home construction had been offered from the earliest days of flight, but home construction had never attained a sustained market. It could even be said that the Wright brothers were the first homebuilders of aircraft.

Most who founded light aircraft firms were classic entrepreneurs, and most were rather youthful when forming their ventures. Early manufacturers were geographically dispersed from the East Coast to Colorado, and those in the East tended to be isolated from those farther west. There were numerous personal associations, however, which led to business collaborations. But disputes, even among friends, were not uncommon, and many ventures broke up over such disputes. One factor very much in the industry's favor was that even during the depression it was recognized that the United States was the only nation in the world with the potential for a mass market in private aircraft, and consequently, for a viable light aircraft industry. Thus new ventures continued to appear in the face of unpromising economic conditions.

2

Growth from the Lindbergh Boom

Government support for aviation development remained extremely limited through the 1920s, but private support for civil aviation began to grow. Among the most important private sources was The Daniel Guggenheim Fund for the Promotion of Aeronautics, Inc. Established in 1925 with a bequest of $2.5 million, the Guggenheim Fund supported many aspects of aeronautics and acted as a spur to private flying and to the industry.[1] Daniel Guggenheim believed, correctly, that public support for aviation and market prospects were strongly linked to flying safety, and the Guggenheim Fund accordingly announced a national Safe Airplane Competition in 1927. The objective was a new design making real advances in flying safety.

In part due to several withdrawals of original entrants, the competition and its $100,000 prize was won by the Curtiss Tanager. A small single-engined cabin model based on the successful Robin, the Tanager was a high-lift biplane employing full-span upper wing slots. It was one of the first short-takeoff-and-landing (STOL) aircraft. Despite its merits, the Tanager proved rather expensive for its time and thus was not practical for production. Also, the prototype was destroyed in 1930 in an airfield grass fire, ending development. The Safe Airplane Competition undoubtedly enhanced design and increased safety awareness, but its measurable impact on the industry was rather minor. The Guggenheim Fund continued to exert an important influence on several other aspects of aviation, however, particularly aeronautical education.

The event which did carry an immediate and major impact was the epochal solo flight of Charles A. Lindbergh from New York to Paris May 20–21, 1927, in his Ryan single-engined monoplane. No event to that point had raised awareness of the potential of aviation more than Lindbergh's feat. Demand for aircraft increased, thousands wished to learn to fly, investors were attracted to aircraft manufacturing, and aviation became regarded as the next growth industry in what was then a booming economy. While new aircraft firms had appeared with regularity to that point, the so-called "Lindbergh Boom," with an expanding market and optimism for its continued growth, attracted scores of light aircraft ventures. Many firms specialized in racing aircraft, as the popular air races became completely civil after 1929, although racers were not candidates for series production. Established firms, including Travel Air with its low-wing Mystery S of 1929, also entered racing.

Personal aircraft market growth in the latter half of the 1920s was sufficient for major manufacturers to employ distributors. In addition, 42 firms used business aircraft by 1928, and business use grew steadily.[2] The importance of aircraft to business was well established by 1930, but business flying required aircraft that were faster and longer-ranged than was typical, leading to larger cabin models such as the Detroiter. Such aircraft could also cost up to $12,000, as compared with the $2,000–$3,000 for small open-cockpit models. Continued growth of fixed-base operations provided operational support for business flying as well.[3] Ford Motor Company was a major business aircraft user, as were most large oil companies, where executives found the speed and flexibility of aircraft valuable in traveling to widely dispersed oil fields. Lieutenant James H. "Jimmy" Doolittle resigned from the Air Corps in 1930 to accept a position as a pilot with Shell Oil Company in St. Louis. Other famed pilots of that era so employed included Frank Hawks at Texaco, Al Williams at Gulf, and Roscoe Turner at Gilmore. Certain corporate-owned aircraft became especially noteworthy in racing.

New Personal Aircraft Entrants

Postwar aircraft production had rebounded from the low of 263 in 1922, of which only 37 were civil, to 6,193 in 1929, of which 5,516 were civil. Sales of new civil aircraft in 1928 exceeded those of military aircraft for the first time since 1916.[4] Market optimism led the formation of new personal aircraft manufacturing ventures to peak in 1928 and 1929, although few were to be of long duration.

One noteworthy entrant was the Sicilian immigrant Giuseppe M. Bellanca, one of the most innovative aircraft designers of the period. Among the earliest experimenters, Bellanca built his first airplane in 1911, and from 1917 to 1920 was an engineer with Maryland Pressed Steel Company. By 1919 he had risen to chief engineer and produced his own designs for the firm's aviation section, but the company failed after the war.[5] Bellanca then met the investor Victor Roos and moved to Omaha, Nebraska, where his Roos-Bellanca cabin monoplane, financed by Roos, won wide attention.[6] The Roos-Bellanca Company, or Omaha Aircraft Company, was active only in 1922 and 1923, however. In 1924, after failing to establish his own firm, Bellanca took a position with Wright Aeronautical Corporation. There he designed the advanced, record-breaking Wright-Bellanca, incidentally, the last aircraft to be produced by Wright Aeronautical.

In early 1927 Bellanca was invited to become president of the new Columbia Aircraft Corporation, founded by the wealthy enthusiast Charles Levine. Columbia had purchased the rights to the Wright-Bellanca from Wright Aeronautical and was interested in building air mail planes. Soon tiring of conflicts with the mercurial Levine, Bellanca left to establish his Bellanca Aircraft Corporation of America on Staten Island, New York, in June 1927. Then the Du Pont family, always interested in aviation, offered to back an aircraft venture, and Bellanca won their support to form a new company in New Castle, Delaware.[7] The new Bellanca Aircraft Corporation was established on December 30, 1927, with 51 percent of the stock owned by Du Pont. The Du Pont family soon experienced policy conflicts with Bellanca, however, and sold its interest.[8] From that point Bellanca designed and produced both military designs and a series of business aircraft, beginning with the six-seat CH-300 Pacemaker of 1929, followed in 1930 by the larger Skyrocket. The CH-300 Pacemaker became a popular bush aircraft as well. While his designs were respected and modestly successful, Bellanca still never won the prestige and fame of others of that era. He also faced periodic challenges from outside investors over his control of the board, although he remained in control.

The Aeronautical Corporation of America (Aeronca) of Cincinnati, Ohio, began operations on November 11, 1928. The Lunken family of Cincinnati, which owned the local airfield, incorporated the new firm and moved into the factory vacated by the defunct Metal

Aircraft Corporation. The firm counted among its directors Robert A. Taft and enjoyed financial support from the politically prominent Taft family. It was the first firm to develop a truly successful light airplane and was to become a major producer of light aircraft.[9] The original C-2 model, designed by Jean A. Roche, who sold the rights to Aeronca, first flew in the summer of 1929. It featured the trademark "bathtub" fuselage and was powered by a small two-cylinder engine also built by Aeronca.[10] The C-2, priced at only $1,495, also brought the cost of flying down to a more reasonable level, important for the later growth of the field.[11] The Aeronca C-2 was followed by the more powerful C-3 in 1931, which became a popular model through the decade.[12] The C-3 was also license-produced in Great Britain during the 1930s. The unrelated C-4 model was designed by company general manager Conrad Deitz, who was killed in an aerial demonstration on September 12, 1931.

In 1927 the brothers C. Gilbert and Gordon Taylor founded Taylor Brothers Aircraft Corporation in Rochester, New York, to produce and sell a small high-wing monoplane named the Chummy, which Gilbert Taylor had designed the previous year. Gilbert Taylor, in common with many contemporaries, was not formally educated but was an instinctive designer. The new firm decided to relocate to Bradford, Pennsylvania, in 1928 in response to economic inducements by that community, and it began operations in February 1929. The Taylor Brothers name was retained even though Gordon Taylor had been killed in a crash.

In its relocation to Bradford the company had attracted the attention and investment interest of several local businessmen, including the oilman William T. Piper, who purchased $400 in Taylor stock and became company treasurer. Although a worthy design, only six Chummy models were sold in the depression-curtailed market, and the company soon faced insolvency. On Piper's recommendation the firm filed for voluntary bankruptcy, after which he purchased the firm's assets for only $761 (incidentally, the only bid), but he still gave Taylor a half interest. The reorganized Taylor Aircraft Corporation emerged in March 1931, with Gilbert Taylor still president, but with Piper effectively in control as treasurer. Piper abandoned his oil business and devoted himself entirely to aviation.

Piper was convinced that a small firm manufacturing simple, low-cost private airplanes could survive even in the depression. Working under that premise, Taylor designed and tested an improved two-seat model, the E-2 Cub, which first flew in February 1931. Piper felt that the E-2 was a superior light aircraft to the Aeronca C-2, and to be competitive set the initial price at only $1,325. Late in 1932 the firm attracted the 19-year-old aspiring designer Walter Jamouneau. Joining the firm in January 1933, initially without salary, Jamouneau refined the Cub design. This resulted in a displeased Gilbert Taylor, who in fact tried to fire Jamouneau over the redesign. The improved Cub was redesignated J-2 in Jamouneau's honor. Piper's support for Jamouneau as well as his aspiration to develop a mass market for the Cub as a trainer for flying schools led his relationship with Taylor, never easy, to deteriorate.[13]

The firm struggled from 1931 through 1935, and Piper sank a major portion of his personal funds into the venture to keep it afloat. But in the latter year the Cub's popularity was such that 228 were sold, constituting almost a quarter of the country's civil aircraft output.[14] More important, the company was approaching profitability after years of losses. Both the Piper and Taylor names were to endure in the field.

A typical short-lived venture of the period was the American Eagle Aircraft Company. Formed originally in 1926 by Edward E. "Ed" Porterfield, a war veteran, in Kansas City, Kansas, it was then incorporated in Delaware in September 1928. Porterfield designed the light, two-seat, high-wing American Eaglet sportplane, similar to the Heath Parasol in presaging later ultralights, to sell for under $1,000. American Eagle built about 500 airplanes, the majority being the A-129 model on which Giuseppe Bellanca assisted, but the company suffered bankruptcy in 1931. Its operations were merged with the Lincoln-Page Aircraft Company of Lincoln, Nebraska, on May 15, 1931, which consolidated operations in Kansas City, Missouri. Ed Porterfield remained as sales manager but left in 1932 to pursue a new venture. Victor Roos, having left Swallow in December 1928 but continuing his involvement with aviation, was president of American Eagle-Lincoln. Unfortunately, this venture also did not endure. Roos later formed the Victor H. Roos Aircraft Company in Kansas City to continue the American Eaglet, but little production ensued. Although cost was the limiting factor in personal aviation, the market still did not favor those aircraft at the smallest, lightest end of the scale.

Aircraft Mechanics, Inc., of Colorado was begun in 1932 by former Alexander employees to succeed the bankrupt Alexander Aircraft Corporation. It attempted to continue the line but failed in 1934. E. M. Laird remained active in the 1930s in racing planes rather than series production.

The Central States Aero Corporation of Davenport, Iowa, was founded in 1926. The designer Don A. Luscombe designed the Monocoupe Model 60 cabin monoplane, which first flew on April 1, 1927, for Central States. Luscombe strongly believed the enclosed cabin represented the future of light aircraft. The firm was succeeded by Mono Aircraft, Inc., in 1929, but Mono Aircraft went into receivership in 1931, and was in turn succeeded by the Monocoupe Corporation, with Luscombe as president. The refined Monocoupe Model 90 two-seat cabin monoplane was a successful design which was to have a long production life. More than 500 of the basic design were produced. Monocoupe was taken over by new owners in 1933 as Lambert Aircraft Corporation in St. Louis, which continued the Model 90. Luscombe resigned in October 1933 to organize a new company under his own name in New Jersey.

R. A. Rearwin, already a successful businessman in Kansas City, Kansas, became interested in aircraft manufacturing after the Lindbergh flight. He formed Rearwin Airplanes, Inc., in May 1929, and began low-volume production of his own designs. Rearwin also produced small aircraft engines under the Ken-Royce nameplate, after his two sons. Struggling in the depression, the firm reverted to a partnership on January 1, 1935, although still owned by the Rearwin family. Rearwin also acquired the established LeBlond Engine Company in 1936.

The new Mid-Continent Aircraft Company of Tulsa, Oklahoma, was purchased by the oilman W. G. Skelly on January 17, 1928. As Mid-Continent, it had produced the C3 biplane trainer and sportplane. Renamed Spartan Aircraft Company on September 27, 1928, it operated the Spartan School of Aeronautics as well as pursuing further aircraft development and production. The C3 was followed by the C3-255 monoplane and C2-60 and C2-65 monoplanes. The C-4 and

C-5 cabin monoplanes were predecessors of the later Executive. Spartan suspended aircraft production by 1935 due to economic conditions, but the School of Aeronautics remained active.

The Cunningham-Hall Aircraft Corporation was organized in Rochester, New York, in 1928 by former associates of the early military firm Thomas-Morse (including Randolph Hall) in association with the automobile firm James Cunningham Son and Co. William T. Thomas, a founder of Thomas-Morse, served as a director and consultant. Cunningham-Hall was an unsuccessful entrant into the Safe Airplane Competition, and market prospects of its promising PT-6 six-passenger all-metal business model disappeared in the depression.[15] It produced a variety of all-metal sportplanes in the mid-1930s but never became a market leader. The company eventually turned exclusively to subcontracting during the Second World War.

It was also from the late-1920s boom period that a more distinct light aircraft engine industry developed. Several light aircraft firms continued to build engines, but specialized engine firms became more prominent. Among the more successful was Continental Motors Corporation of Michigan, founded in 1905, which first built an aircraft engine in 1928. Its flat-four, or horizontally-opposed, configuration became a standard for small engines. The Lycoming Manufacturing Company of Pennsylvania, founded in 1910, also developed its first aircraft engine in 1928. Production began in 1931, and it also became a leader in the field. The aircraft engine operation, after being acquired by E. L. Cord, became the Lycoming Division of the Cord-controlled Aviation Corporation in a reorganization on January 1, 1936.

The Jacobs Aircraft Engine Company of Pottstown, Pennsylvania, founded in 1929 by Philadelphia businessmen backing the engineer Al Jacobs, specialized in small radial engines. The Warner Aircraft Corporation of Detroit first tested the small Scarab engine in 1927 and developed it and the larger Super Scarab into the 1930s. Fairchild's Ranger Aircraft Engine Division, extending from its purchase of Caminez in 1925, was brought within the new Fairchild Engine and Airplane Corporation organized in 1936. Ranger specialized in the in-line design, which powered a series of personal aircraft and military trainers. Kinner Motors, Inc., of Glendale,

California, was organized in 1939 after the bankruptcy of the older Kinner Airplane and Motor Corporation in 1938, and it became the largest engine producer on the West Coast. The Aircooled Motors Corporation took over the activities of the old Franklin Automobile Corporation and continued to develop Franklin aircraft engines, which first entered service in 1938. Somewhat larger radial engines widely used on business aircraft were the Pratt & Whitney R-985, developed from the original Wasp, with 450–600 hp, and the Wright Whirlwind, with 300–450 hp.

The Autogiro

One approach in the quest for an airplane for everyman was the autogiro, combining certain conventional fixed-wing and vertical takeoff, rotary-wing concepts. The autogiro involved a free-turning rotor for lift and a conventional tractor engine and propeller for forward thrust. The combination of ease of operation, safety, and no need for long runways made the autogiro the potential solution to mass-market aerial transportation.

The autogiro in the United States will always be linked to Harold Pitcairn. Pitcairn, an aviation enthusiast and scion of a wealthy Philadelphia family, began his Pitcairn Aviation flying school and passenger service on November 2, 1924, and expanded into aircraft construction in 1926.[16] Pitcairn possessed extensive industry experience prior to his involvement with the autogiro, having first worked as an apprentice to Curtiss in 1914. He met Agnew E. Larsen, later a close associate, at flight school in 1916.[17] Initially, Pitcairn was interested in producing aircraft only for his own needs, as he expanded into air mail in addition to passenger service. He placed the engineer Larsen, with previous experience at Thomas-Morse, in charge of his Pitcairn Aircraft subsidiary on June 17, 1927. He achieved success first with sportplanes and then with mailplanes, culminating in the famous Pitcairn Mailwing of 1927. He received his first air mail contract on January 28, 1927.

The Spanish designer Juan de la Cierva, working in England, had discovered the autorotation principle, essential to autogiro operation and also valuable eventually in helicopter development. The term *autogiro* (or autogyro) originally was proprietary to Cierva but became generic. Pitcairn and Larsen first made contact with Cierva in 1925, and by the late 1920s the potential of the autogiro had attracted widespread interest. Pitcairn passionately believed in its promise of becoming the affordable, easy-to-fly, mass-market aircraft.

Pitcairn transformed his manufacturing arm into Pitcairn Aircraft Company in 1928, continuing in mailplanes as well as holding the U.S. license for the autogiro. He developed the Cierva C-8 model and tested it successfully on December 19, 1928. He then organized the Pitcairn-Cierva Autogiro Company of America in February 1929 to hold the Cierva license and to control the patents. Although suffering a factory fire in November 1929, Pitcairn still completed his first PCA-1 autogiro the next month. He moved to his new Willow Grove factory in suburban Philadelphia in 1930, and by that time was the leading proponent of the autogiro in the country. He improved on the original Cierva concept by angling propeller thrust toward the rotorblades, enabling the autogiro to attain nearly vertical takeoff and hovering capability.

Pitcairn-Cierva Autogiro Company became the Autogiro Company of America (ACA) in January 1931. The company aggressively developed and marketed autogiros, including demonstrations by Amelia Earhart, who in fact set an autogiro altitude record of 18,415 feet on March 6, 1931, at Willow Grove. Pitcairn Aircraft Company remained as the operating company under ACA. Despite winning the Collier Trophy in 1930 for his accomplishment with the autogiro, Pitcairn never succeeded in gaining wide market acceptance or practical application.

Pitcairn still produced mailplanes, but their day was rapidly drawing to an end. He also produced the PA-7S three-seat sportplane version of the Mailwing from 1930, but still regarded the autogiro as the mass-market aircraft that might even begin to replace the automobile. The PA-18 Tandem two-seat open-cockpit model of 1932, at a base price of $5,000, was marketed as the Model T of the air, but orders were few. The enclosed cabin model PA-32 and the still larger 4-5 place PA-19 cabin model also appeared in 1932. Another problem was that the autogiro was never as safe as orginally claimed, with mishaps and damage occurring frequently. Although Pitcairn fell far short of his mass-market goal, his efforts aided the later development of practical helicopters. The name of the original manufacturing arm, Pitcairn Aircraft, was changed to Pitcairn Autogiro Company in January 1933, after Mailwing production had ended.[18]

Kellett Autogiro Company, also in the Philadelphia area, was the second American licensee, purchasing the license from ACA. Formed by the brothers W. Wallace and Rodney G. Kellett, the firm followed Pitcairn in developing and seeking practical applications for autogiros, and Wallace Kellett and Harold Pitcairn were longtime friends. Also participating in the firm were the Ludington brothers, involved in air mail operations, and Elliott Daland, a founder of Huff Daland Airplanes. The Englishman W. Lawrence LePage served as chief engineer. Kellett models generally were wingless, while Pitcairn models were configured with short wings. A few examples of the Kellett autogiro were sold to the Army, but commercial sales proved disappointing, and the company later turned to development of helicopters.

Depression Era Struggles

Another outgrowth of the Lindbergh Boom, beyond broad expansion of aircraft markets, was the formation of large aviation holding companies, principally through mergers of existing firms. While primarily oriented toward the military and commercial markets, the holding companies also had some involvement with private aviation. With the depression the Lindbergh Boom proved to be a bubble, however, and most larger firms abandoned their personal aviation interests.

The first holding company effort, the Detroit Aircraft Corporation of 1927, acquired the well-known Ryan and Lockheed firms, plus several others primarily in private aviation, but went bankrupt and disappeared in the 1931 depression. United Aircraft and Transport Corporation (UATC) of Connecticut, organized in February 1929, gained an involvement in personal aircraft, although of rather brief duration, through its acquisition of Stearman. Lloyd Stearman, while an outstanding pilot, designer, and engineer, was less effective as an executive, and soon encountering differences with UATC management, he resigned in December 1930. UATC phased out production of his designs in 1931 with the market decline.

The Stearman Division of Boeing formed in 1934 from the UATC Stearman operations had no managerial or design participation by Lloyd Stearman.

The Aviation Corporation (AVCO) also became involved with personal aviation, first with Fairchild and from 1933 with Stinson. Shortly after its organization in 1929, largely by Wall Street interests, AVCO acquired 55 percent of Fairchild Aviation Corporation stock, bringing with it Fairchild's Farmingdale factory. Sherman Fairchild became a vice president of AVCO. He withdrew in disenchantment with AVCO management on April 1, 1931, however, leaving his aircraft operations but regaining his 55 percent stock holding. Then when E. L. Cord gained complete control of the Aviation Corporation in March 1933, he brought it under his Cord Corporation and with it the Vultee and Stinson aircraft operations and Lycoming engine operations. The manufacturing divisions remained at their existing locations as they were combined under the subsidiary Aviation Manufacturing Corporation, headquartered in Chicago, in 1934. The former Fairchild aircraft models were phased out.

Curtiss Aeroplane and Motor, later Curtiss-Wright, built a substantial and long-term private aviation involvement. Particularly noteworthy was its association with Robertson Aircraft Corporation of St. Louis, formed on July 1, 1923, primarily as a mail carrier by the brothers Frank H. and Major William B. Robertson, a wartime aviator.[19] Among the firm's mail pilots was Charles A. Lindbergh. William Robertson's contact with Curtiss concerning his interest in small aircraft resulted in the three- to four-seat Robin high-winged cabin monoplane, a Curtiss-Robertson collaboration built in the Curtiss factory at Garden City, Long Island. A Curtiss-Robertson Manufacturing Corporation, 50 percent owned by Curtiss, was formed on November 9, 1927, with capital of $500,000.

The Robin first flew in the spring of 1928, but Curtiss's heavy military involvement at the time made Robin production uneconomical. The company adopted Robertson's suggestion to produce the Robin in St. Louis, and a new factory was constructed. Entering production on August 7, 1928, initially with the 90 hp Curtiss OX-5 engine and later with the 170 hp Curtiss Challenger, the Robin became one of the most successful light aircraft of the time, with a total of 769 produced and with many record-breaking flights to its credit.[20] Other

Curtiss-Robertson designs included the Thrush and the Kingbird, a twin-engined light transport, but neither was produced in quantity. The Robertson Aircraft Corporation remained a separate entity.

The merger of Curtiss A and M and Wright Aeronautical forming the Curtiss-Wright Corporation was achieved on August 29, 1929. Curtiss-Robertson had been fully merged into Curtiss earlier, on June 26, 1929. Then Travel Air, initially under Wright Aeronautical, was absorbed into Curtiss-Robertson in 1930.[21] The ultralight CW-1 Junior two-seat sportplane, with a design contribution by Walter Beech, followed, with some 270 produced from 1930 to 1932. Beech, already president of Curtiss-Robertson, became president of Curtiss-Wright Airplane Company when it was renamed, still located in St. Louis and specializing in civil aircraft. Ralph Damon and William Robertson were vice presidents, although Robertson stepped aside in 1933 after completion of the merger.[22] All Travel Air production was consolidated at the St. Louis factory and the Wichita plant was closed. The St. Louis company soon introduced the larger Challenger to complement the existing product line, but the Model 16 Sport, the last aircraft built under the Travel Air nameplate, ended in 1933 due to the depression. The Curtiss Speedwing, a fast open-cockpit model powered by the Wright Whirlwind and originally designed under Travel Air, survived until 1936.

Another Curtiss subsidiary, the Moth Aircraft Company, had been organized in St. Louis in 1928 to produce the de Havilland Moth trainer under license. Production was rather brief, however, and Moth was soon reorganized into the Curtiss-Wright Airplane Company under the Curtiss-Wright holding company.

Cessna by 1929 had grown to rank with Travel Air, Swallow, and Stearman as one of the Wichita "Big Four" in light aircraft, out of sixteen aircraft manufacturers located in the city. A new factory also was opened in 1929. There were rumors that Cessna would follow Travel Air and Stearman in joining a large holding company, a factor of no small concern to Wichita, but Clyde Cessna chose to remain independent. The young firm suffered after the October market crash, however, and Cessna struggled for control. Bankers installed a new president over him, and the company was finally forced by the bankers and other creditors to shut down production indefinitely on January 31, 1931. The firm never declared bankruptcy, but its production future was very much in doubt.[23] After being forced out of his company, Cessna immediately started a new C. V. Cessna Aircraft Company with his son Eldon, but more for racing aircraft than mass production, and this venture did not enjoy a long life.

Walter H. Beech, uncomfortable in a large corporation, resigned as a vice president of Curtiss-Wright Corporation and as president of Curtiss-Wright Airplane (Curtiss-Robertson) in March 1932, determined to return to Wichita and establish his own company. With his wife Olive Ann Beech he formed Beech Aircraft Company on April 1, 1932, with capital of $25,000.[24] Most regarded the venture as foolhardy at the time: 1932 was the poorest year for civil aircraft production and demand since the 1922–25 period. Joining Beech was his Travel Air associate T. A. "Ted" Wells, who undertook design of the Model 17, a large, powerful, and fast executive biplane popularly known as the Staggerwing. Beech began production in 1933 with space leased in the then-inactive Cessna factory, with only a handful of employees. The Staggerwing faced a depressed market, but Beech managed to sell 18 in 1933. Beech then was able to move into the old Travel Air factory in 1934, and sales continued to build gradually.

The engineer Dean B. Hammond purchased the rights to the Parks Aircraft Division of Detroit Aircraft Corporation after its collapse, and Parks formed the basis for his Hammond Aircraft Corporation on May 11, 1932.[25] He also bought the rights to the Ryan Speedster sportplane from Detroit. Major development effort, however, was with the innovative twin-boomed pusher-engined Model Y, supported by a Bureau of Air Commerce contract. Hammond, a pioneer of tricycle landing gear, later developed the design further.

When Sherman Fairchild severed his ties with the Aviation Corporation in 1931, he moved his headquarters to Hagerstown, Maryland, location of the original Kreider-Reisner firm. Production of Kreider-Reisner designs had been suspended in the market collapse of 1930, but Fairchild's Kreider-Reisner Division reinstated production of the F-22 two-seat open-cockpit biplane. The most important new design was the F-24 three- to four-seat cabin monoplane developed in 1932. The F-24 in succeeding versions became a long-term production success for both civil and military markets.[26] In 1934 Sherman

Fairchild formed a successor Fairchild Aircraft Corporation to control all manufacturing, and the Kreider-Reisner name disappeared.

Curtiss-Wright reorganized on divisional lines, and the Curtiss-Wright Airplane Division was created as an equal division with Curtiss Aircraft at Buffalo and Wright Aeronautical. It then was renamed the St. Louis Division and focused primarily on commercial aircraft as Robin production declined. Its major effort was the Condor airliner of 1932, of which 44 commercial and military models were produced.[27] While not a major success, the Condor at least kept the St. Louis line open. Curtiss still devoted some effort to personal aircraft development, resulting in its more modern Models 19L and 19W Coupe low-wing all-metal two- and three-seat cabin monoplanes. The CW-19R tandem-seat trainer was a more powerful development.

Ed Porterfield reentered the industry in August 1934 by forming Porterfield Aircraft Corporation, still in Kansas City, Kansas. Famed pilot Roscoe Turner was associated. Porterfield acquired the design and production rights for a light aircraft, the Wyandotte Pup, in exchange for stock in his company, and began producing a line of improved Porterfield trainers, beginning with the Model 35-70 (for the year and engine horsepower). He operated in this capacity until transforming to war production in 1942.[28]

# # # # #

All aircraft manufacturers suffered greatly during the depression, but the economic impact was more severe and immediate for the personal sector than for the military sector. Military procurement continued until 1932–1933 from earlier appropriations, but the personal aircraft market disappeared almost overnight. Remaining firms struggled to develop models that could sell in a limited market, but they rarely achieved profitability.

Growth of the business of small aircraft constructed from plans or component kits, despite potentially lower costs, was severely constrained by certification and safety concerns. Those alternatives had long existed, but the government offered little encouragement to the field. Homebuilts faced major difficulties in gaining Approved

Type Certificates and, consequently, government registration numbers. Many amateur-built aircraft were soundly designed and constructed, of course, but others were not, heightening safety concerns. Further, the priority of the Department of Commerce at the time was commercial service, and there were fears that publicity over crashes of amateur-built light aircraft would detract from public confidence in commercial air safety.[29] One model, however, achieved some success, the Pietenpohl Air Camper. Designed by B. H. Pietenpohl of Minnesota in 1934, the Air Camper was a simple two-seat design with a high parasol wing, and it proved popular with amateur builders during the late 1930s.

The dream of an affordable, mass-produced airplane for the public remained. Eugene L. Vidal, appointed Director of the Aeronautics Branch in 1933, was determined to achieve that goal. He noted that in some 15 years of private flying development there were still fewer than 7,000 registered private airplanes, and felt that flying could become far better established if only small airplanes could be made more affordable.[30] In December 1933 he offered $500,000 in grants for development of a two-seat "poor man's airplane," with a target price of only $700. The craft was to be all metal, as it was felt that metal would enhance public confidence in its safety and durability.[31] The "$700 airplane" concept attracted great interest and publicity, but established producers, aware of the economic realities, generally declined to compete, leaving the field to experimenters. Many firms also opposed the program as detrimental to the sector, in that potential customers might postpone their purchases in expectation of lower prices, further damaging their sales prospects.[32]

The grant later was withdrawn and the goal recognized as unrealistic.[33] The Aeronautics Branch became the Bureau of Air Commerce on July 1, 1934, and a successor development program at that time saw the appearance in 1936 of the tricycle-geared Arrowplane of Waldo Waterman, predecessor of the first roadable airplane, the advanced Hammond Y-1, a roadable autogiro, and other innovative designs. Vidal left his post in 1937 with his efforts having achieved no permanent results and the average personal airplane still costing in the $2,000 range. Prices could be as low as $1,500, but they ascended to more than $25,000 for the largest twin-engined business models.

Whatever the overall economic conditions, the private aircraft market was still limited by cost, both for purchase and operation. The resulting high turnover in personal aircraft ownership and availability of good used aircraft also limited the demand for new models. But a major market expansion lay just ahead.

3

Market Expansion and War Production

A more modern aircraft industry structure had evolved by 1935, due both to market development and new legislation and regulation. Technical progress also had been steady despite weak demand, inadequate government support, and shaky industry finances. Within the new structure, aircraft and engine manufacturing operations were largely separated, and those commercial airlines previously owned by holding companies or manufacturers were spun off as independent firms. The aerial service term had narrowed to such functions as aerial photography, cropdusting, weather reporting, and air taxi work. Business and personal flying gained in prominence and became classified separately from aerial service. In addition, light or personal aircraft production by 1935 had become a largely separate and distinct sector of the industry, and numerous personal aircraft producers sought a niche in the limited market.

Design progress had been significant, as monoplanes increasingly displaced biplanes and enclosed cabins succeeded open-cockpit models. Some producers moved toward all-metal structures, and engines and instrumentation became more efficient and reliable. Larger military and commercial firms such as Curtiss-Wright retained some activity in private aircraft, but firms specializing in the field dominated the market. Aeronca, Piper, Taylor, and Luscombe, primarily producing two-seaters, emerged as the Big Four of the light aircraft field by the end of the decade. WACO, Bellanca, and the later leaders Cessna and Beech focused on larger personal aircraft models, with lower volume. The following survey traces the more significant developments.

Personal Aircraft Firm Survey

The Piper-controlled Taylor Aircraft Corporation continued to expand with its successful Cub. The company finally attained profitability in 1936, with 515 examples of its J-2 Cub being sold.[1] Increasing friction between William T. Piper and Gilbert Taylor led to a management breakup in that year, however. Piper insisted on keeping prices as low as possible, not the priority of Taylor. Taylor also felt himself increasingly undermined and isolated by Piper. Piper confronted Taylor with an offer to purchase his interest for $5,000, which Taylor felt obligated to accept. Piper then took over as chairman and president while remaining treasurer. The improved J-3 Cub appeared in 1937.

Gilbert Taylor was determined to remain a factor in the industry and moved to Butler, Pennsylvania, early in 1936, where he established Taylorcraft Aviation Company. He felt his Model A, with side-by-side seating, would be superior to the otherwise similar J-3 Cub. Needing more production space, Taylor relocated to Alliance, Ohio, on July 8, 1936, and entered direct competition with Piper.[2] The Model A was in production by 1937, but Gilbert Taylor became concerned about a possible Piper lawsuit over use of the Taylor name, although none materialized.

Finances had been precarious, but the company became better capitalized in April 1937 by the participation of local engineer and businessman W. C. Young. The firm was renamed the Taylor-Young Airplane Company, with Young as vice president. A public stock offering followed, strongly supported by the city of Alliance, wishing to retain what was viewed as a strong growth company.[3] Although Young soon ended his association, apparently late in 1937, the firm name remained unchanged at the time.

The firms appeared to be on a sounder base, but Gilbert Taylor realized that he needed to offer new models in the increasingly competitive market. First, the more powerful Model 50 appeared. Sales were encouraging, but a fire and explosion at the factory on August 23, 1938, further strained finances. Rescue came on October 19 from Fairchild Aviation Corporation, which provided a $30,000 loan in exchange for 50,000 shares of Taylor stock.[4] Fairchild also installed Richard H. Depew as vice president and general manager, as Gilbert Taylor again lost managerial control of a company bearing his name.* The company was renamed Taylorcraft Aviation Corporation in 1939 and progressed as a market competitor.

Another significant development was the award of a production license to Taylorcraft Aeroplanes, Limited, in Great Britain, a company established by British owners on November 21, 1939, for that purpose. Taylorcraft thus became one of the earliest American aircraft companies to license its production abroad. The British Taylorcraft firm produced adaptations of Taylorcraft designs during the war for artillery spotter duties.

The original Piper-operated Taylor factory in Bradford was completely destroyed by fire on March 17, 1937. The high flammability of light aircraft construction materials meant that insurance was very costly, and W. T. Piper was grossly underinsured for the loss. He reallocated funds earmarked for expansion to restart production in a temporary factory. But the disaster led Piper to consider relocation, and he received offers from as far away as Oklahoma. He settled, however, on the town of Lock Haven, some distance east of Bradford. Already a man of near-legendary frugality, Piper was not known for his high wages or tolerance of unions, but these views did not pose a problem in Lock Haven, and local business interests assisted in the move. Attracting generous local financing and reestablishing production in an old silk mill, the venture progressed rapidly. Overcoming the disruption of the fire and the move, the Cub continued as the most popular aircraft in America, with 687 produced in 1937. Renamed Piper Aircraft Corporation in November 1937, the firm by then employed the three Piper sons.

William T. Piper still was faced with significant financial challenges. Production startup expenses in Lock Haven, in addition to the uninsured loss of $200,000, caused a substantial drain, but he succeeded with a $250,000 public stock offering on March 3, 1938, followed soon by an additional $750,000 stock sale.[5] A Cub nonstop Newark-Miami round-trip flight during May 17–20, 1938, using crude in-flight refueling and covering 2,420 miles, greatly enhanced public recognition of the reliability of light aircraft and boosted sales. In 1939 1,806 Cubs were sold, and Piper lowered the price of the basic Cub trainer to $995, the first mass production aircraft ever priced

* Fairchild Aviation was primarily involved with photography and survey operations, while manufacturing was under the separate Fairchild Engine and Airplane Corporation.

under $1,000. W. T. Piper was becoming known as the Henry Ford of the aircraft industry.

Finally on the verge of real prosperity, Piper earned an after-tax profit of $157,823 for the 1940 fiscal year. By the end of 1941, 10,000 Cubs had been built, a record at that time for a basic design, and the Cub name had become almost generic for a light aircraft. Piper also introduced the J-4 Coupe, with side-by-side seating, in 1939, followed by the J-5B Cruiser in 1940, its first three-seat aircraft, which was to serve as the basis for Piper's postwar developments.

Don Luscombe located in Trenton, New Jersey, in 1934 for his new venture, attracting sufficient financial backing and enjoying access to a plentiful labor supply.[6] The Luscombe Airplane Development Company began operating in leased factory space in February 1935, initiating production of the luxurious all-metal Phantom, which had won certification in August 1934. The Phantom possessed many advanced features and was one of the first light aircraft built with extensive subcontracting. Its $6,000 price was too high for the market of the day, however, as was that for the smaller Model 90, limiting sales. Luscombe thereupon designed the two-seat, 50-horsepower Model 8 in 1937. Named the Silvaire, it sold for only $2,000 and gained immediate popularity.

The company was required to sell stock for production capital and went public as Luscombe Airplane Corporation in 1938. The major investor was the millionaire Leopold Klotz, originally from Austria, who then began to intervene in operations. Early in 1939 Klotz imposed a new manager on the company, sparking protests from the staff loyal to Don Luscombe.[7] Luscombe thereupon sold his interest to Klotz in April 1939 and resigned as president and director. Although only 44, Luscombe ended his involvement in the industry. The company sold 1,200 Silvaires from 1938 to Pearl Harbor and was a major subcontractor on Pitcairn's autogiro production, but private aircraft production was suspended at the outset of the war due to metal shortages.

In 1937 Aeronautical Corporation of America (Aeronca) ended production of the veteran C-3 in favor of the more modern Model K Scout, with side-by-side seating. The Model L low-wing two-seater was introduced in 1935, while the high-wing Scout was joined by the more powerful Chief. The Taft family, after sustaining considerable losses, sold Aeronca for $750,000 to Walter J. Friedlander in 1935. He installed his son Carl as manager and relocated production from Cincinnati to nearby Middletown. Jean Roche left in the mid-1930s, but Carl Friedlander proved an astute manager, and the newer Aeronca models provided the strongest competition to the Piper Cub until the Second World War.

Spartan, determined to reenter manufacturing, designed the noteworthy Executive, a powerful five-seat luxury business transport which first flew on January 1, 1936. Executive production began in 1937, and though built in small numbers, it gained wide recognition in the late 1930s and was adopted by the military after the outbreak of war as well. Spartan also designed the NS-1 military primary trainer, which was ordered by the Navy as the NP-1.

Pitcairn Autogiro reorganized as Pitcairn-Larsen Autogiro Company in 1940, reflecting Agnew Larsen's increased role. Larsen had long desired more active management participation, and Harold Pitcairn already was stretched as a manager. Pitcairn and Larsen soon developed major conflicts over costs, however, precipitating Larsen's departure.[8] The firm maintained low-volume autogiro production, although principally of its PA-33 and PA-34 models for the Army and Navy rather than for the civil market. The company was renamed AGA Aviation Corporation (for autogiros, gliders, and airplanes) in 1941, with Pitcairn taking a new partner.[9]

Beech progressed as Model 17 Staggerwing sales totaled 36 in 1935. The firm was rechartered in Delaware as Beech Aircraft Corporation on September 16, 1936, with a stock capitalization of $100,000. The company adopted the marketing name Beechcraft for its products. Its speed and comfort enabled the Staggerwing to gain a substantial market in the late 1930s, and 424 had been produced by the beginning of the war. Model 17 developments then were employed in numerous military roles. On January 6, 1937, Beech was able to repurchase the Travel Air Wichita factory from Curtiss-Wright for $150,000. Beech also tested the high-performance Model 18 "Twin Beech" on January 15, 1937, which began to find a prewar market as the top-line business transport. Beech employment reached 250, and the company issued its first public stock offering also in 1937.[10] The "Twin Beech" was to be produced in large numbers in several wartime military versions.

In May 1933 Clyde Cessna's 22-year-old nephew Dwane Wallace, a new engineering graduate of Wichita University, took a position with Beech Aircraft Company. He worked on Staggerwing developments under Ted Wells but resigned at the end of the year to help his uncle regain control of his company. Aided by his older brother Dwight, a lawyer, Dwane Wallace prevailed in a proxy fight with the banking interests and gained control of Cessna on January 10, 1934. Clyde Cessna again became president, and the firm resumed limited production after the three-year hiatus. Cessna and Wallace set out to design a plane which could sell in the depression market, resulting in the C-34 (for the year), which continued Cessna's successful cantilevered high-wing concept. Powered by a Warner Super Scarab engine and selling in the $5,000 range, the C-34 led to a series of progressive designs. Owing to the firm's shaky finances, Wallace found it necessary to forgo a salary during this period and asked Eldon Cessna to do the same. Eldon declined and moved to California for other pursuits.

Clyde Cessna's management role eventually declined in his revived company; he had been deeply affected by seeing a friend die in a crash. Although only 57, he sold his interest to his nephews and officially retired on October 8, 1936.[11] While Cessna's retirement was described as voluntary, Eldon Cessna maintained in later years that his father had been forced aside by his Wallace nephews.[12] Dwane Wallace succeeded his uncle as president and was in sole control of the company. After building 42 C-34s, Wallace brought out the Model C-37 in 1937, which led to the classic Model C-38 Airmaster. Fifty C-37s were sold in 1937, giving the firm a strong market foothold, and a total of 186 Airmasters of all series sold through 1941. Wallace then developed the Model T-50 Bobcat twin-engined transport/trainer in 1939. Only 30 civil Model T-50s were sold before the war, but large-scale production followed as Cessna gained major military orders for the Bobcat.

Lloyd Stearman, after an unhappy tenure as president of the revived Lockheed Aircraft Corporation, joined the Bureau of Air Commerce in January 1935. While serving there, he assisted Dean Hammond with development of his Model Y. He resigned late in 1936 to join Hammond in organizing the Stearman-Hammond Aircraft Company in San Francisco. The refined Model Y, designated Y-125, entered production, but the company encountered financial difficulties and ended production at only 15 aircraft.[13] Stearman left in 1937 for other aviation-related ventures, but Hammond continued as a subcontractor with the company under his name.

Henry A. Berliner, after the sale of his Berliner-Joyce military firm to North American Aviation, incorporated the Engineering and Research Corporation (ERCO) in Maryland in 1930 to produce various types of machinery used in aircraft production. The engineer Fred E. Weick, who had developed the NACA streamlined cowl in 1928, resigned from NACA and joined ERCO in 1936 as chief designer as the firm moved into aircraft.

Weick had designed the small W-1 experimental model in 1934, featuring tricycle gear, a pusher engine, and a high parasol-type wing, in response to the Department of Commerce program. With ERCO, Weick redesigned the W-1 into the low-winged, tricycle-geared, tractor-engined W-1A (originally ERCO 310), which first flew in October 1937 and became the famous Ercoupe.[14] The Ercoupe was noteworthy for its lack of foot pedals, being flown entirely with the control wheel, and it was spinproof. Entering the market in 1940, only 112 Ercoupes were completed before production was interrupted by the war, but the Ercoupe became a classic light aircraft.

The Curtiss-Wright St. Louis Division won a Bureau of Air Commerce contract to develop the advanced CW-25 Coupe, a twin-engined all-metal design which interested the agency as appropriate for advancing private flying. Adapting the wing of the earlier CW-19 Coupe, the CW-25 Coupe prototype was delivered to the Department of Commerce in 1936, but the design did not reach the civil market. It was to be the last Curtiss effort in private aviation, but was produced in substantial numbers as the wartime AT-9 Jeep. Conceptually, it presaged the postwar generation of light twin-engined business transports.

The Stinson Division of AVCO continued to produce larger cabin monoplanes, principally the classic Reliant, introduced in 1934 and developed from the original Stinson designs. Initially produced with a straight wing, the Reliant soon adopted the better-known gullwing. Stinson also developed the Model 10 lightplane, designed by Lewis E. Reisner, a founder of Kreider-Reisner. It first flew on February 13, 1939. The developed Model 10A became the long-produced

Voyager. Some 500 were produced before Pearl Harbor, and many were acquired by the Army as the L-9.[15]

In a related development, Vultee Aircraft of California, another division of AVCO and gaining in stature as a military aircraft builder, acquired the small Barkley-Grow Aircraft Company of Detroit early in 1940. Barkley-Grow had been founded in 1936 by Archibald S. Barkley, whose career dated back to Wright and Curtiss, and Commander Harold B. Grow, with a staff composed largely of former Stinson personnel. Barkley-Grow designed the twin-engined T8P-1 executive transport. Although a sound design, it did not find a wide market, given the competition, principally the "Twin Beech." Then in August 1940, the Stinson Division was combined as Vultee-Stinson under AVCO, which absorbed Barkley-Grow the following month.

Bellanca aircraft continued their strong reputation into the 1930s, with many record-breaking flights to their credit. The Pacemaker and Skyrocket were sound designs but did not find commercial success. Nor did their successor, the larger Aircruiser, aerodynamically a sesquiplane with airfoil-section struts. Bellanca also developed military designs, but did not gain orders. The firm tended to rely on proven concepts and features which became outmoded with the rapid progress of the late 1930s, and sales were declining even before war requirements forced a production suspension.[16] Bellanca did, however, build the small three-seat 14-9 Cruisair in 1939, a modern low-wing retractable-gear design which would be revived after the war.

Under Clayton J. Brukner, Waco remained a leading private aircraft manufacturer through the 1930s. Chief designer Francis Arcier, an Englishman and design pioneer, joined Waco in 1930 and enjoyed one of the strongest reputations in the industry. The Model F biplane, introduced in 1930, was a popular design, as were the C series and S series. The model CSO tandem two-seat biplane was even adapted as a single-seat light fighter in 1931 and ordered by Brazil, although serving only briefly in a combat role. The N series of 1937, a four-seat cabin biplane with tricycle gear evolved from the C series, was unique, and was produced in small numbers. The last prewar taildragger model was the E series, produced until 1940. The firm found that its designs were becoming rapidly outdated by the end of the decade, however, and with the approach of war converted to military production.

Grumman, while primarily a military contractor, embarked on limited diversification into private aviation with the G-21 Goose twin-engined amphibian transport, which first flew on May 29, 1937.[17] The Goose won market acceptance in its narrow segment, and was joined by the smaller G-44 Widgeon amphibian, which first flew on June 28, 1940. Grumman amphibians served in the Second World War, and the company was to continue in the smaller civil amphibian field after the war.

Lockheed, growing as a military and commercial airliner manufacturer, also considered a stake in private aviation during 1937. Lockheed President Robert Gross first offered Mac Short, then managing the Boeing Stearman Division, the management of a private aircraft subsidiary.[18] Short declined at the time. There were subsequent discussions between Gross and Victor Emanuel, who controlled the Aviation Corporation, about selling his three manufacturing arms, Vultee, Stinson, and Lycoming, with the expectation that Lockheed would merge Lycoming with its Menasco engine operation into a single engine company. Stinson then would be moved to the Northrop El Segundo factory, which Gross proposed to acquire from Douglas.[19] This proposal, which would have given Southern California a strong stake in the personal aircraft field, likewise did not reach fruition.

Lockheed instead established a new private aircraft subsidiary, AiRover Aircraft, located adjacent to its main Burbank factory. It was incorporated on August 17, 1937. Mac Short then joined Lockheed as manager of AiRover. It developed the innovative Starliner executive aircraft, with twin engines coupled to drive a single propeller, but did not proceed with production given dim market prospects at the time. AiRover was renamed Vega Airplane Company on June 1, 1938, and turned to military work.

Fairchild Engine and Airplane Corporation was formed on November 10, 1936, to control all Fairchild aircraft and engine production. The advanced Fairchild F-45 low-wing high-performance five-seat business aircraft of 1936, somewhat competitive with the Executive, was another success, although built only in small numbers. While continuing personal aircraft, primarily the F-24, the Aircraft Division became increasingly military oriented, winning a major primary trainer competition for the Army in 1939.

There were further entrants into the light aircraft field in the period, including revivals of older firms. Despite its daunting economics, the field still held a strong attraction for many. Among these, Lambert Aircraft Corporation of St. Louis, which had developed the two-seat low-wing Monosport, sold rights to the Dart Manufacturing Corporation of Columbus, Ohio, in 1937. The Monosport design evolved into the Dart Model G and was produced in small numbers. Then Culver Aircraft Corporation, founded by K. K. Culver in 1939 in Columbus, with Al Mooney as chief designer, took over the rights to the Dart Model G. With a design contribution by Mooney, it became the Culver Cadet and entered production in 1940 after the firm moved to Wichita. In November 1941 Culver was bought by Walter H. Beech and Charles G. Yankey, a Beech associate since the Travel Air days. Yankey became the firm's president. Monocoupe Corporation was revived in Robertson, Missouri, in 1937 and resumed low-volume production of the Model 90 high-wing two-seat design until the wartime suspension in 1942. It also tested the Monocoach, a four-seat business transport design with twin 90 hp engines. One of the first light twin-engined cabin monoplanes, it did not move into production.

Reuel T. Call, in association with his uncle Ivan Call, his brother Spencer Call, and others, established Call Aircraft Company in Afton, Wyoming, in 1939 to produce his rugged light aircraft design. The first Call Model A flew in 1941, but the outbreak of war meant access to needed materials was almost impossible, and the civil market also was disrupted. Further, the firm's isolated location precluded its winning any military subcontracts. Call thus survived on repair work and by building propeller-driven snow cars for mountain winter transportation. Active aircraft production had to await the end of the war.[20] The firm is noteworthy, however, because at over 6,000 feet elevation its factory was the highest in the world and its products advanced the art of mountain aviation. The location proved to be a distinct advantage in testing aircraft under adverse conditions.[21]

The Bennett Aircraft Corporation of Fort Worth, Texas, was formed in 1939 and designed an experimental twin-engined light transport, but it soon moved toward development of light aircraft of bonded plywood construction. With reorganization in 1941 the company became Globe Aircraft Corporation, still in Fort Worth. The company's first design, the low-wing two-seat GC-1 Swift, held promise, but production was forced to await the end of the war.

The Rearwin Company was reincorporated as Rearwin Aircraft and Engines, Inc., in 1937. The Models 6000 and 6000M Speedster light monoplanes from 1935 continued, as did the Model 7000 Sportster. The somewhat larger three-seat Model 8135 Cloudster appeared in 1940. The more modern Skyranger also was developed in 1940, but all aircraft production ended in that year as the firm transformed entirely to engines.

The designer Ben O. "Benny" Howard, who won the Bendix Trophy in 1935 with his Mr. Mulligan high-winged racer, established Howard Aircraft Corporation in Chicago on January 1, 1937, to produce his designs. His DGA series of cabin monoplanes, beginning with DGA-8, extending from the Mr. Mulligan, reached the civil market. The series also saw extensive service in wartime as the UC-70, and many were produced for the Navy as the DG 1-3 Nightingale.

The Interstate Aircraft and Engineering Corporation was formed in California in April 1937, initially manufacturing components and subassemblies for the aircraft industry. Located in the original Northrop El Segundo hangar, Interstate benefited from its proximity to North American, Douglas, and Lockheed. It produced its new two-seat S-1B Cadet in 1940. The Army acquired 250 Cadets as the L-6, and eight civil models were later taken over as the L-8.

Max Harlow, an engineer with experience at Northrop, Douglas, Lockheed, and Hughes, founded the Harlow Engineering Corporation, later Harlow Aircraft Company, in California in 1939 to manufacture a light all-metal executive aircraft, the PJC-2. The PJC-2 was rather similar to the Spartan Executive, although smaller. Quantities of the PJC-2 were taken over by the Army Air Forces for war service as the UC-80, but in 1941 the company transformed its production over to subcontracting.

Swallow Airplane, descended from the original Laird firm of 1920, reentered aircraft production in 1937. It produced a two-seat trainer in 1940, but little series production was achieved and the Swallow name disappeared.

Meyers Aircraft Company was founded by the 27-year-old engineer Allen H. Meyers in Tecumseh, Michigan, in 1936. Its Model OTW-160 biplane trainer of 1939 was widely used during the war by contract training schools and would continue in production postwar.

After the demise of the Ryan nameplate with the Detroit bankruptcy, the pioneer aviator T. Claude Ryan of *The Spirit of Saint Louis* fame formed a new Ryan School of Aeronautics on June 5, 1931, still in San Diego. Deciding to enter manufacturing again, he then formed Ryan Aeronautical Corporation on May 26, 1934. The flying school became a subsidiary.[22] Ryan concentrated on military trainers, but the two-seat open-cockpit all-metal S-T (for Sport-Trainer), which first flew on June 8, 1934, also reached the civil market. Then the three-seat all-metal SCW-145 low-wing cabin model of 1937, with advanced features later seen in postwar aircraft, gained a small market share from 1938 through 1941. Ryan, with others, then turned entirely to military trainer production and flying school operations.

# # # # #

While the United States remained the only nation with a mass-market potential in personal aviation, the numbers of privately owned aircraft and licensed pilots were miniscule compared with those for automobiles. A 1938 survey stated that personal aviation was in no sense a mass market, there being only some 10,000 privately owned airplanes, compared with millions of automobiles.[23] The field had nonetheless developed significantly during the economically troubled 1930s (see Table 3-1). The four years before Pearl Harbor brought major growth, and the Civil Aeronautics Board (CAB) estimated the private aircraft fleet at 24,000 in 1941. The pre-Lindbergh total probably was fewer than 2,000. The Aircraft Owners and Pilots Association (AOPA) was founded in May 1939 and expanded steadily.

Another major development resulted from the new Civil Aeronautics Authority (CAA), established on August 22, 1938. The CAA, which replaced the Bureau of Aeronautics of the Department of Commerce, operated autonomously. The CAA began a nationwide Civilian Pilot Training Program (CPTP) in 1939. Numerous flying schools across the country received CPTP contracts, and the program began

Table 3-1
Total Civil Aircraft,* 1927–1941

As of December 31	Number
1927	2,740
1928	5,104
1929	9,922
1930	9,818
1931	10,680
1932	10,324
1933	9,284
1934	8,322
1935	9,072
1936	9,229
1937	10,836
1938	11,159
1939	13,772
1940	17,928
1941	26,013

* Includes commercial airliners, since records of the day did not break out personal or business aircraft. Commercial airliners did not number more than a few hundred during this period, however, and it may be concluded that most of the civil fleet consisted of personal or business aircraft.
SOURCE: Civil Aeronautics Administration, *Statistical Handbook of Civil Aviation,* 1950, p. 25.

training thousands of new pilots, increasing both aircraft demand and the growth of flight schools (see Table 3-2). Not incidentally, the program also facilitated the approaching wartime buildup.

Design progress, an improving economy, and airways and airfield development combined to strengthen personal and business aviation. Numerous manufacturers, including the light aircraft Big Four; growing firms such as Beech and Cessna, plus WACO, ERCO, and Stinson; and smaller producers such as Culver, Howard, Monocoupe, Rearwin, and Spartan, made the field vibrant. Wichita, however, did not dominate production as it had earlier. Manufacturing became widely dispersed across the country, and Beech and Cessna were then low-volume producers. Wichita aircraft industry employment was only some 2,000 in 1939. Consumers had enjoyed

Table 3-2
Number of Certified Planes of Five Seats and Under

As of January 1, 1939	Number
Aeronca	853
Beech	163
Cessna	114
Stinson (Vultee)	779
Curtiss-Wright	741
Fairchild	567
Fleet	191
Luscombe	61
Piper	1,658
Rearwin	170
Taylorcraft	623
Monocoupe	293
Waco	1,050
Total	7,412

SOURCE: John H. Geisse, *Report to W. A. M. Burden on Postwar Outlook for Private Flying,* GPO, Washington, 1944, p. 80.

a broad choice, but the outbreak of war precluded further design development, and all civil production was suspended for the duration. Most manufacturers became subcontractors to major firms or converted to other war-related production.

General Aviation Manufacturers in Wartime

As the Air Corps was reorganized into the Army Air Forces (AAF) on its journey toward equal status with the Army and Navy as a military service, organic Army aviation was reborn on June 6, 1942. The roles of battlefield observation and artillery spotting, previously performed by large, heavy observation aircraft with O-designations, were gradually supplanted by light aircraft originally designed for the civil market.

William T. Piper first wrote Secretary of War Henry Stimson on February 18, 1941, proposing light aircraft for such military roles.[24] After some resistance, the Army began placing orders for light aircraft such as the Cub. John E. P. Morgan, a Piper director, lobbied for the change on behalf of Aeronca and Taylorcraft as well as Piper and became an unsalaried liaison between the manufacturers and the War Department.[25] Tests of the Cub during large-scale Army maneuvers of 1941 were promising and gained the strong support of Third Army Chief of Staff Colonel Dwight D. Eisenhower. With civil aircraft production suspended, orders for L-series light military aircraft enabled the industry to remain in aircraft production. The O-designation was abolished. Piper, Aeronca, Taylor, Stinson, Beech, Cessna, and smaller firms filled large orders for liaison and training aircraft of their design.

The primarily private aircraft firms contributed to the wartime aircraft production effort in other ways as well. In addition to adapting civil models to military roles, they produced military designs under license and engaged in subcontracting and component production for military producers. Some were active in all.

The Stinson Division of the merged (from 1943) Consolidated Vultee Aircraft Corporation produced more than 12,000 light aircraft during the war, including over 3,000 L-5 (originally O-62) Sentinels, developed from the Voyager, plus several hundred AT-19 Reliants for the U.S. forces and for the Royal Navy under Lend-Lease. The L-1 Vigilant, originally designated O-49, was built in the new Vultee factory in Nashville, Tennessee, which opened in 1940 and offered a lower-cost labor force. Inspired by the German Fieseler Storch, the Vigilant still was not a major success and was soon succeeded by the L-5 Sentinel.[26] The Wayne, Michigan, factory remained active with production of the veteran Reliant and large numbers of the Voyager and L-5.

Beech, which had become a major business aircraft firm before the outbreak of war, transformed into a significant military aircraft producer. Walter Beech missed extensive work time due to health problems in 1940, and his wife became increasingly active in management. At a time when Walter was hospitalized after a stroke, and Olive Ann was also hospitalized giving birth to a daughter, certain Beech directors attempted to take over management. Olive Ann Beech immediately foiled the attempt and remained in control. She negotiated a $50 million loan from a consortium of banks, then a Reconstruction Finance Corporation (RFC) loan of $13.5 million later in 1940, to finance a major expansion.[27]

Beech undertook a major subassembly operation from 1943 in supplying wing sets for the A-26 advanced attack aircraft to the Douglas factory in Tulsa, Oklahoma, where the A-26 was assembled and not far from Wichita. Unlike most light aircraft firms, however, Beech designed and tested a combat aircraft, the advanced XA-38 Grizzly. The XA-38 was a sound design worthy of production, but the powerful Wright R-3350 engines of the XA-38 were urgently needed for the B-29 program and could not be diverted for the new design.[28] Beech produced 5,257 of its Model D-18 "Twin Beech" during the war under such designations as C-45, AT-7, and AT-11 for the AAF, plus Navy and allied variants. It also built 1,771 of the AT-10 Wichita, a smaller twin trainer constructed of wood due to then-anticipated shortages of strategic materials.

Cessna achieved a major success with an order for its Model T-50 twin-engined light transport from the Royal Canadian Air Force in 1939 as the Crane trainer. Dwane Wallace negotiated a $500,000 line of credit from a Wichita bank to finance the start of production. Canada became the center of British Empire flying training, and the T-50 played a key role in that undertaking. Although not to the scale of its neighbor Beech, Cessna expanded rapidly with war contracts and opened a new factory in nearby Hutchinson, Kansas, in 1942. Sales reached $13.6 million and profits $1.6 million for 1941. Production of the T-50, including prewar civil models and Canadian orders, totaled 5,402 through the war, most under the AT-8, AT-17, and UC-78 designations. Both Cessna and Beech unionized in 1940, following unionization by most larger aircraft firms. Cessna built Waco CG-4A gliders under license, and also developed the C-106 Loadmaster cargo transport in 1943. But the C-106, while demonstrating promise, simply lost out in the battle for materiel priorities. Like Beech, Cessna subcontracted on the Douglas A-26 and was an important subcontractor to Boeing-Wichita as well.[29]

From 1941 through 1944 Taylorcraft produced its L-2 Grasshopper spotter, or liaison aircraft, a tandem-seater developed from the civil Model D. The company also built a number of small military glider trainers derived from the L-2. Upon completion of production contracts, Taylorcraft engaged in subcontracting for the duration of the war.[30] Gilbert Taylor was effectively removed from management in 1942, as Fairchild's influence increased. Fairchild then sold its interest in the spring of 1943 to outside financiers, as the troubled ownership and financial history of the firm continued.

The Piper J-3 Cub, by far the most numerous light aircraft, received the first large military order, for 1,500, on February 10, 1942. The most successful of the military light aircraft types, it was produced throughout the war as the L-4 Grasshopper (all light liaison aircraft were named Grasshopper) as well as in a Marine Corps and a Navy ambulance version, a total of 6,028 being delivered. In common with most aircraft producers, Piper unionized in 1941, although this displeased William Piper.

After Luscombe ended civil aircraft production and switched over to subcontracting in 1942, the FBI removed Leopold Klotz from his position in the firm due to his foreign nationality. But Klotz gained U.S. citizenship in 1944 and regained control of his firm on June 6, D-Day.[31] He immediately began planning a move from New Jersey to Texas for postwar production.

Bellanca suspended civil aircraft production in 1941 and undertook military subcontracting. It also completed 39 Fairchild AT-21 Gunner combat trainers under license. Globe's wartime activity included production of 600 Beech AT-10 trainers and subcontract work. The Swift, for the postwar civil market, first flew in January 1945.

Waco devoted its efforts to war production from August 1941, but its civil models were not adopted in large numbers by the military. Its most important contribution was in design and manufacture of troop-carrying gliders. Waco built more than 1,600 of its glider designs, beginning with the CG-3. The widely used CG-4A was also manufactured in large numbers under license by several firms. William Robertson, whose Robertson Aircraft Corporation was a glider subcontractor, died along with several prominent citizens in a glider crash on August 1, 1943. His company exited at the end of the war.

Aeronautical Corporation of America changed its name to Aeronca Aircraft Corporation in 1941. The two-seat civil Model 65 Super Chief, developed from the earlier Model 40 Chief, was modified into the military L-3 (originally O-58) Grasshopper, and more than 1,400 were produced. Aeronca produced Fairchild PT-19 and PT-23 primary trainers under license as well.

Among smaller general firms, Rearwin was sold in October 1942 to a group of New York investors and reconstituted as Commonwealth Aircraft, Inc. Headquarters remained in Kansas City, and activity was devoted to war production, including Waco gliders. The oilman J. Paul Getty gained ownership of Spartan in 1942, and the company remained active in training wartime pilots. Howard Aircraft Corporation's DGA-15 cabin monoplane was ordered by the military, and the firm also produced the Fairchild PT-23 under license, but all government contracts were canceled in mid-1944 and the company exited. In 1945 Harlow, another wartime subcontractor, purchased rights to the neighboring Interstate firm's designs and production equipment after the latter abandoned aircraft production. Culver was noteworthy for its series of radio-controlled target drones, based on its aircraft design. The Army had developed a requirement for drones to simulate enemy air attacks, and Culver became one of the more successful producers of drones.

# # # # #

Light aircraft manufacturers contributed greatly to the war production effort, and their cooperation with the major military manufacturers would serve them well for the future. But their focus remained on private flying. Companies either offered updated prewar designs or developed completely new designs reflecting aeronautical advances as they eagerly anticipated a postwar private flying boom.

Walter Beech, early 1920s. (*Source: Ablah Library, Department of Special Collections, Wichita State University.*)

Travel Air biplane, late 1920s. (*Source: Ablah Library, Department of Special Collections, Wichita State University.*)

Stearman C-3R sales brochure photo, circa 1927. (*Source: National Air and Space Museum, Smithsonian Institution.*)

New York Aeronautical Salon, May 1930. (*Source: National Air and Space Museum, Smithsonian Institution.*)

Taylor factory, Bradford, Pennsylvania, circa 1930. (*Source: National Air and Space Museum, Smithsonian Institution.*)

Travel Air production workshop, 1929. (*Source: Ablah Library, Department of Special Collections, Wichita State University.*)

Travel Air factory, circa 1930. (*Source: Ablah Library, Department of Special Collections, Wichita State University.*)

Luscombe 8A Silvaire production at Trenton, New Jersey, late 1930s. (*Source: National Air and Space Museum, Smithsonian Institution.*)

Beech Staggerwing, the first modern executive aircraft, 1930s. (*Source: Ablah Library, Department of Special Collections, Wichita State University.*)

First prototype Beech Model 18, January 15, 1937. (*Source: Ablah Library, Department of Special Collections, Wichita State University.*)

Cessna Airmaster production, circa 1939. A Model T50 Bobcat is seen on the right. (*Source: National Air and Space Museum, Smithsonian Institution.*)

WACO factory, 1930s. (*Source: National Air and Space Museum, Smithsonian Institution.*)

A double production milestone: The 10,000th Piper Cub and the 5,000th Lycoming engine for the Cub. W. T. Piper and C. O. Samuelson of Lycoming exchange congratulations, Lock Haven, Pennsylvania, 1941. (*Source: National Air and Space Museum, Smithsonian Institution.*)

Walter and Olive Ann Beech inspecting SNJ production line during Second World War. (*Source: Ablah Library, Department of Special Collections, Wichita State University.*)

Beech AT-10 production line, Second World War. (*Source: Ablah Library, Department of Special Collections, Wichita State University.*)

Ercoupe production, Riverdale, Maryland, late 1940s. (*Source: National Air and Space Museum, Smithsonian Institution.*)

Cessna factory, early 1950s. (*Source: Ablah Library, Department of Special Collections, Wichita State University.*)

The Bell Model 47 in use by the San Francisco Police Department, late 1940s. (*Source: Bell Helicopter Textron.*)

4

The Postwar Era, 1946–1954

The War Production Board lifted the suspension on civil aircraft production on May 17, 1945, and personal aircraft firms eagerly resumed civil production. Established prewar firms were joined by new entrants seeking to participate in the forecast private flying boom. Yet stability was to remain elusive. Several firms were quickly forced from what rapidly became an overcrowded market, and others sought or were forced into merger for survival. Even the successful Beech and Cessna firms, Wichita neighbors, conducted serious merger discussions in June 1945, though both had prospered during the war and were optimistic about the future. Walter Beech and Dwane Wallace could find few points of agreement, however, and talks were ended.[1] Years later, neither company would comment publicly about the negotiations.[2]

Large military aircraft firms, which had experienced massive contract cancellations with the end of the war, faced diminished future military aircraft prospects as well. With the general recognition that there simply were too many companies for anticipated business, some explored merger, with the support of or even pressure from the government. But many also sought to exploit the anticipated personal aircraft boom, posing a potential threat to the oft-beleaguered personal aircraft sector. The following account traces the complexities of the postwar general aviation sector, as well as its technical progress.

The General Aviation Upheaval

While the military sector reached a low point early in 1946, production and employment rose sharply in personal aircraft. Similar to the automobile market, there was a virtual flood of new light aircraft to meet pent-up demand.

Despite widespread parts shortages, a total of 35,000 civil aircraft was produced in 1946 (see Tables 4-1 and 4-2), of which 31,594 were small personal models led by such firms as Aeronca, Cessna, Piper, and Taylorcraft.[3] Beech had not yet placed its new design into production, and its output trailed temporarily. In the same year, only 1,059 new military aircraft were delivered.[4]

General aviation as it now exists, broadly defined as all aviation outside the military and scheduled commercial sectors, was born in the immediate postwar period. But forecasting the future of private aviation at the time was in some ways as complex as for the military market. Private aviation, given depression and war, was recognized as having been a highly volatile business over the past two decades.[5] But AOPA membership exceeded 40,000 in 1947, and the number of licensed private pilots approached 250,000. The Aircraft Industries Association (AIA), successor to the Aeronautical Chamber

Table 4-1
General Aviation Unit Sales, 1946*

Piper	7,780
Aeronca	7,555
Cessna	3,959
Taylorcraft	2,483
Ercoupe	2,503
Luscombe	2,483
Stinson	1,436
Globe (Swift)	1,054
Temco	563
Beech	299
Bellanca	288
Republic	196
Funk	194
Navion (NAA)	146
Total	30,939

*Other sources also report some 400 Culver Model V sales.

Table 4-2
Civil Aircraft

Total Registered Civil Aircraft, Selected Years		Summary by Number of Engines, November 1, 1947	
December 31, 1927	2,740	Single engine	80,537
December 31, 1929	9,922	Two engine	3,447
December 31, 1932	10,324	Three engine	48
December 31, 1934	8,322	Four engine	484
December 31, 1936	9,229	Unspecified	7,441
December 31, 1938	11,159	Gliders	673
July 15, 1941	22,354	Lighter-than-air	14
December 31, 1945	37,789		
December 31, 1946	81,002		
November 1, 1947	92,644	Total	92,644

Civil Aircraft Registration by Major Manufacturers as of November 1, 1947

	Manufacturer	Number
1.	Piper	19,007
2.	Aeronca	12,456
3.	Taylorcraft	8,075
4.	Cessna	7,242
5.	Fairchild	5,796
6.	Consolidated Vultee*	5,094
7.	E. & R. Corp. (ERCO)	4,217
8.	Boeing**	3,899
9.	Luscombe	3,945
10.	North American	1,915
11.	Douglas***	1,613
12.	Waco	1,622
13.	Beech	1,484
14.	Globe	1,078
15.	Curtiss-Wright	827
16.	Lockheed***	394
	All others	13,980
	Total	92,644

*Includes Stinson.
**Includes Stearman.
***Principally commercial airliners.
SOURCE: *World Aviation Annual, 1948,* p. 279.

of Commerce, established both a Private Aircraft Council (PAC) and a Helicopter Council to address those segments. The Civil Aeronautics Administration, heretofore strongly oriented toward scheduled commercial service, began to be more supportive of nonscheduled and personal flying. John H. Geisse promoted personal flying for the CAA.

Postwar optimism was such that there were many forecasts that a true mass market would arrive. A figure of 500,000 personal aircraft by 1950 was bandied about.[6] One market survey reported some 78,000 sales prospects for light aircraft, and that 119,000 individuals were interested in personal helicopters. Even the U.S. Department of Commerce developed a forecast of 200,000 personal aircraft a year.[7] In addition, civilian flight training was covered by the G. I. Bill, enabling thousands more to gain private licenses. Factors felt to limit demand, however, were safety concerns, costs, regulations, and airfield availability.[8]

Light aircraft use by the military had become firmly established during the war, and demand continued in the postwar era as large numbers of military variants of the Super Cub, Champion, Navion, and others were ordered. The military also supported the sector by ordering specialized liaison and trainer designs, which general aviation firms often could supply more efficiently than large military firms. And small helicopters originally designed for military use were easily adaptable to civil uses, supporting development of a civil market. The influential Finletter Commission report of 1948, while primarily concerned with the military and national security aspects of air power, recognized the role of the personal aircraft industry, and recommended its preservation and support. NACA, while still focused on military research, held a conference in the summer of 1946 to disseminate wartime research applicable to the design of personal aircraft.[9] Thus the light aircraft sector (see Table 4-3), while lacking the political influence of larger industries, still gained significant support from the military, NACA, CAA, Commerce, and other agencies.

Personal aircraft remained only a small percentage of the total value of aircraft production, but most large military aircraft firms considered entering the field, regarding it as a potential offset to the postwar decline of military production.[10] Military firms also anticipated

Table 4-3
Nine Major Personal/Light Commercial Airframe Companies, 1947*

1. Aeronca Aircraft Corp.
2. Beech Aircraft Corp.
3. Bellanca Aircraft Corp.
4. Cessna Aircraft Corp.
5. Engineering & Research Corp.
6. Luscombe Airplane Corp.
7. Piper Aircraft Corp.
8. Taylorcraft, Inc.**
9. Texas Engineering & Manufacturing Corp.

SOURCE: *Survival in the Air Age: A Report to the President's Air Policy Commission,**** GPO, Washington, D.C., 1948, p. 51.
*The Stinson Division of Consolidated Vultee was not included.
**Bankrupt end of 1946.
***Popularly known as the Finletter Commission.

a private flying boom and felt that their greater resources, including their engineering and production expertise, could be applied to advantage in that market. While a large-scale entry into the sector by military firms would have seriously threatened established light aircraft manufacturers, it developed that of the numerous personal aircraft studies and prototypes undertaken by the larger firms, few entered production.

Kaiser Industries briefly considered producing an all-metal four-seat design at its Fleetwings operation in Bristol, Pennsylvania, but declining market prospects ended the idea. Goodyear, the rubber producer, but also a major wartime aircraft manufacturer, tested the small GA-2 amphibian in 1947 but decided against production. Even Rohr, a major military subcontractor, developed a personal airplane, a small V-tailed tricycle-geared model, but likewise did not proceed with production.[11]

Douglas, reviving the name of its original airplane, tested the innovative five-seat pusher-engined Cloudster in 1946. The Cloudster was of advanced configuration, with twin rear-mounted engines geared to a single pusher propeller, but it remained experimental.

Lockheed developed the small Little Dipper and Big Dipper personal aircraft, maintaining its practice of astronomical names. Lockheed decided both were unpromising and wrote off the developments. The single-seat Little Dipper, functionally a flying motorcycle, also was proposed to the Army but attracted no interest. Lockheed's 14-seat Saturn commuter airliner was tested in 1946, but at a projected price of some $100,000, it fell victim to availability of military surplus C-47s at around $25,000 each.

Consolidated Vultee (Convair) developed several new personal aircraft models, seeking to continue a strong market presence with more modern successors to the Voyager. The Model 106 Skycoach, an all-metal, pusher-engined, four-seat design of the Stinson Division, flew in April 1946. The twin-engined Model 116 followed in July 1946. The Model 111 Air-Car, designed by the engineer T. P. Hall, explored the roadable airplane or flying automobile concept, and was followed by the more powerful Model 118 ConVAirCar, which first flew on November 15, 1947. With rising costs, efforts soon ended for all new models, however, and Convair's only general aviation entry remained the Stinson Voyager. Hall pursued the Air-Car independently after Convair dropped development but eventually ended his efforts.

Grumman's entry into the personal aircraft market, planned before the end of the war, was to offer nothing for its future. Grumman had developed and tested both the small G-63 Kitten I landplane and the G-65 Tadpole amphibian in 1944. Then in 1946 came the G-72 Kitten II, with tricycle gear and somewhat similar to the Ercoupe. Assessing the market and concerned about a price structure somewhat higher than the competition, Grumman wisely decided not to enter the civil market at the time. The smaller Widgeon and the new, larger Mallard twin-engined amphibians also found limited markets, a major factor being the availability of cheap surplus amphibians. Only 59 Mallards and 76 Widgeons were built, but their development at least kept Grumman in the amphibian field.[12] Grumman was to reenter the personal aircraft field later.

Fairchild chief executive J. Carlton Ward determined that the company, in addition to military work, should maintain a major presence in the personal aviation field. He pushed for production of the new F-47 low-wing four-seat business aircraft and established the Fairchild Personal Planes Division in Kansas late in 1945, developments opposed by major stockholder Sherman Fairchild. When Fairchild regained control of his company in 1948, precipitating Ward's departure, the division was disbanded. The Glenn L. Martin Company, also concerned over future military aircraft prospects, conducted a series of design studies for personal aircraft, but none reached the prototype stage.

The only personal aircraft developed by military firms to enter production were the North American Navion and the Republic Seabee. The four-seat low-wing NA-143 Navion (a rough acronym of North American Aviation) first flew on January 16, 1946. The Navion drew on the design of the P-51 Mustang fighter and had considerable customer appeal in the luxury segment of the market, and NAA pursued production in the field whereas Lockheed, Douglas, and Convair did not. The Navion also won Army orders as the L-17 liaison aircraft. While North American had regarded the civil market as a necessary diversification move, initially strong demand for the Navion disappeared when the market for new light aircraft suddenly dried up in 1947, resulting in a loss of $8 million on the venture.[13] North American thereupon stopped Navion production on April 15, 1947, at 1,108 aircraft, selling all rights to the smaller Ryan Aeronautical Company on June 25. Ryan, which had not continued its prewar civil models, felt that the Navion would enable it to compete in the postwar market.

After purchasing the rights from North American, Ryan produced the Navion until 1951, including further L-17 military models. But after building some 1,200 Navions for both civil and military markets, Ryan ended production due to the press of Korean War military requirements and concentrated on subcontracting.[14] The Navion, however, would have a long development and production life with other firms, including twin-engined conversions.

Republic's RC-1 four-seat pusher-engined light amphibian first flew in November 1944. The developed RC-3 Seabee entered production in 1946 for the civil market, but postwar inflation forced costs to the point that it was no longer the economy model first envisaged. Underpriced originally at $3,995, the price soon rose to over $6,000, but losses still increased. Although selling well initially, Seabee production ended with the general market contraction of 1947. A further complication was Republic's purchase in December 1945 of Aircooled Motors, maker of the Seabee's Franklin engine. Although intended to enhance Republic's cost control, the purchase instead added pressure to its already strained finances.[15] After the Seabee ended production, Aircooled was resold to the ill-fated Tucker Corporation in March 1948, but the Seabee episode was a factor in a Republic management shakeup.

Whatever the potential advantages of military firms in light aircraft, the early postwar experience served to validate the existence of a personal aircraft manufacturing sector. That distinct sector in fact continued, although many firms would later become controlled by larger corporations, both of the aircraft and other industries.

In the heady immediate postwar period, established private aircraft manufacturers felt that they could market small models for as little as $2,000, and larger four-seat models which could sell for $4,000, but rising costs and market uncertainties made these figures unattainable.[16] The major demand in the short term was for proven prewar designs, such as the Champion, Cub, Taylorcraft, Silvaire, Voyager, Swift, and Fairchild F-24, but certain manufacturers, especially Beech and Cessna, readied completely new designs.

It soon became apparent, however, that the industry had seriously overproduced in 1946. Flying schools became saturated. Then the general collapse in new aircraft demand and the industry's inability to provide aircraft at low prices sealed the fate of all but a handful of manufacturers and models (see Table 4-4). Personal aircraft deliveries declined to only 7,037 in 1948.[17] New designs generally cost in the $5,000–$9,000 range, too expensive for major market growth. Disposals of war surplus light aircraft by the RFC were another constraint on new aircraft demand. Yet even as numerous firms exited light aircraft production, new competitors entered, and a major development in the 1949–1953 period was the appearance of light twin-engine business aircraft, which would become a major growth segment.

A new investor, Nash Russ of Detroit, bought control of Taylorcraft on May 13, 1945. The company still experienced financial difficulties, however, exacerbated by overexpansion and the expense of developing a four-seat model, and creditors forced a shutdown on

Table 4-4
Annual Production and Sales, General Aviation Aircraft

Year	Units	Factory sales ($ millions)
1946	35,000	$110.0
1947	15,594	57.9
1948	7,037	32.4
1949	3,405	17.7
1950	3,386	19.1
1951	2,302	16.8
1952	3,058	26.8
1953	3,788	34.4
1954	3,071	43.4

NOTE: Figures account for fixed-wing aircraft only.
SOURCE: GAMA, *General Aviation Statistical Databook* (1990–1991 ed.), p. 4.

November 8, 1946. Taylorcraft declared bankruptcy in December 1946, and its assets were placed up for sale.[18] Gilbert Taylor, along with Ben J. Mauro, a former distributor, reentered the industry by forming Taylorcraft, Inc., on April 25, 1947, purchasing the rights and leasing part of the Alliance, Ohio, factory for limited production. With new financing, the firm relocated in 1949 to the Beaver County Airport in Pennsylvania for development of new production as well as for subcontracting work.[19] The four-seat Model 15 Tourist finally entered production in 1950, followed by the Model 19 Sportsman, developed from the 1939 Model B. The firm was actively managed by Mauro on behalf of Taylor. With the benefit of additional financing secured in 1951, the firm would attain modest success during the decade.

Waco, another major prewar firm which had fallen behind the leaders, made a final attempt at a competitive model for the postwar business market with its unusual pusher-engined Aristocraft, which first flew in December 1946. The firm decided market prospects did not look promising, and production was never undertaken. Waco then withdrew from the aircraft field on June 2, 1947, but continued production of aircraft parts and small machinery and subcontracting.

Aeronca enjoyed initially strong postwar sales of the Model 7 Champion, developed from prewar models, and of the L-16 Army version. It introduced the side-by-side Model 11 Chief and refined Super Chief, also extending from its prewar designs, and the four-seat Model 15 Sedan in 1947. It then developed the experimental low-wing Arrow. Over 10,000 postwar Champions had been built when all production was suspended in 1951 in the face of weak demand. Design rights were sold to a new Champion Aircraft of Wisconsin, which later resumed production. Aeronca remained a significant component producer and subcontractor.

The renamed Monocoupe Engine and Airplane Corporation was acquired by new investors and relocated to Melbourne, Florida. It resumed production of its prewar two-seat Model 90 design during 1947 but ended production in 1949. Harlow exited in 1946. Max Harlow, with partners, reentered the field in 1947 as Atlas Aircraft Company. The modern H-10 four-seat design was tested but did not enter production.

Two new investors bought Culver from Charles Yankey on November 14, 1945. Culver entered the postwar market with its Model V (for Victory), developed from the prewar Model G, but suffered bankruptcy later in 1946. Cunningham-Hall, out of the airframe business since before the war, was dissolved in 1948, happily with no loss to stockholders.[20] Spartan tested the modernized Model 12 Executive with tricycle gear in 1946, but production was not undertaken and the company exited aircraft production. The tricycle-geared Ercoupe was produced in some numbers until 1950, when production ended in the generally weak market.

Commonwealth Aircraft, successor to Rearwin, acquired Columbia Aircraft Corporation of Valley Stream, Long Island, a wartime contractor, early in 1946. Commonwealth closed its Kansas plant and consolidated operations at Valley Stream, but Columbia retained its separate identity. The new Commonwealth Trimmer light twin-engined amphibian introduced in 1946 was not produced, but the Skyranger 185, developed from the prewar Rearwin design, reentered the market. Limited production was suspended on March 14, 1947, and the firm exited aircraft.

Among smaller manufacturers, Meyers attempted to compete with the modern low-wing MAC 125, but still with taildragger landing gear. It was followed by the more powerful tricycle-geared MAC 145 and by the retractable-geared MAC 145T, produced from 1948 until 1956, when production was suspended. Call produced the small CallAir A-2 and A-3 strut-braced low-wing models in limited numbers during this period. The rugged Call models enjoyed a reputation in the western states but still encountered difficulty competing against the industry giants.

Bellanca, despite respected designs in the four-seat Cruisair and the larger Cruisemaster, introduced in April 1949, simply lost out in postwar market competition against such firms as Piper, Cessna, and Beech, and suspended production in 1952. All patents, rights, and production equipment for the Cruisemaster were sold to Northern Aircraft, Inc., of Alexandria, Minnesota, in 1955.

Luscombe moved to Garland, Texas, in 1945 to continue production of the updated Silvaire and the larger four-seat Sedan. The company considered for a time a twin-boom pusher-engined personal aircraft developed from the prewar Stearman-Hammond design, but it did not proceed. The Silvaire then faded in competition with the new Cessna 120 and 140 models. Ironically, Cessna had been inspired by the Silvaire in developing its postwar line. By 1949 Luscombe could not meet its payroll, and efforts to secure an RFC loan were unsuccessful.[21] The firm declared bankruptcy, and Leopold Klotz withdrew and returned to New York. The company soon reorganized under a new Luscombe Holding Company, which held 85 percent of Luscombe Airplane.

J. B. Baumann, a founder of Mercury Aircraft and later with Lockheed, founded Baumann Aircraft Corporation in 1945 to develop his B-250 Brigadier twin pusher-engined executive aircraft. The Brigadier first flew in June 1947, and during 1949 Piper, interested in the business market, considered and then rejected acquiring Baumann and its production rights. The more powerful B-290 was tested in 1951, but with no financing, the Brigadier unfortunately never attained production.

The engineers Ben M. Anderson and Marvin Greenwood had started a small company in 1941 for personal aircraft development. Both joined Boeing in Wichita for the war, but revived the firm afterwards in Houston, Texas. Anderson, Greenwood and Co. produced five of its unusual AG-14 light twin-boom pusher design, the first flying on October 1, 1947, but soon moved into subcontracting.

A sidelight to the anticipated postwar private flying boom was the recurring dream of a truly practical roadable airplane. While the concept was not new—Harold Pitcairn had given consideration to a roadable autogiro—M. B. "Molt" Taylor, a former naval aviator, developed the concept more extensively than had anyone previously. An important feature of his design was that the flying surfaces could be towed behind the automobile section, thereby conferring greater flexibility in operation. Taylor formed Aerocar, Inc., in February 1948, and the Aerocar I first flew in October 1949. Taylor developed his Aerocar for years, but a market for such a craft never developed. Other roadable designs, the prewar Waterman Arrowbile and the postwar Fulton Airphibian and ConVAirCar, were short-lived.

Even with the major market reverse of 1947, a general aviation Big Three of Beech, Cessna, and Piper emerged. Their status was enhanced and protected somewhat by the nonentry and exits from the private aircraft market by larger military manufacturers, but long-term survival was more attributable to design quality and competitiveness.

Cessna moved quickly after the end of the war to establish design leadership. It led the industry in promoting the "family car of the air" concept.[22] With others, Cessna was strained by the severe market contraction in 1947, even diversifying into metal furniture and industrial products briefly. The Hutchinson factory, closed at the end of the war, transformed into industrial products manufacture. Most significant, however, was Cessna's rapid conversion to all-metal aircraft, ending all production of fabric-covered airplanes in 1949. Cessna also retained the high-wing configuration, although strut-braced rather than fully cantilevered. The two-seat Model 120 and refined Model 140, designed early in 1945, achieved instant success and were followed in 1948 by the four-place 170. The Model 170, at a base price of $4,995, was more expensive than the Piper Pacer and Aeronca Sedan, but its all-metal construction and greater power conveyed strong advantages.

The larger radial-engined Models 190 and 195, similar in configuration to the prewar Airmaster, were produced from 1947 to 1954. The more modern Model 180, with the same wing as the Model 170 but with a larger fuselage and more powerful engine, appeared in 1953.

The Models 170 and 180 and their developments became market leaders, and Dwane Wallace became recognized for his marketing acumen. Dwight Wallace, however, left in a dispute with his brother, never publicly discussed, and joined Beech. Cessna successfully entered the light twin-engined business market with the Model 310, which first flew on January 3, 1953.

Cessna also became established as a military aircraft contractor and with expanding business operated three plants in the Wichita area from 1951. The L-19 Bird Dog Army liaison aircraft, developed from the Model 170, won a spirited contract competition in June 1950 for a successor to early liaison models. The L-19 was widely used in Korea and after, with production totaling 2,460 by 1959. Later, Cessna won the Air Force design competition for a jet-powered basic trainer. The prototype Cessna XT-37 first flew on October 12, 1954, and entered large-scale production. Cessna also participated in the Korean War expansion by manufacturing major components for the Boeing B-47, also produced in Wichita, thus continuing its wartime subcontracting relationship.

Beech, which built a strong military production record during the war, returned immediately to the civil market. The refined G-17S postwar model of the classic Staggerwing ended production in 1948 after 90 were delivered. Total production since 1932 was 781. Critical for the future was the new Model 35 Bonanza, with its distinctive V-tail, which first flew on December 22, 1945. In 1947, its first full year of production, 1,229 Bonanzas were sold, a strong counter to the overall weak demand of that year. The Bonanza, in one sense a successor to the Staggerwing, gained a commanding position in the higher-powered luxury four-place segment, despite being the most expensive of the field with a base price of $7,975. With continuous improvements the Bonanza was produced for decades. During March 6–8, 1949, a Bonanza fitted with wingtip tanks and piloted by William P. Odum made a solo nonstop flight of 5,273 statute miles from Honolulu to Teterboro, New Jersey, a record for a private aircraft design. The flight took 36 hours. Odum, unfortunately, died in a racing crash later that year.

Beech also tested the innovative Model 34 "Twin Quad" commuter airliner in 1947. Although a promising design, which paired four 375 hp Lycoming engines buried in the wings to drive twin propellers and possessed the V-tail configuration of the Bonanza, Beech decided market prospects were not encouraging and ceased development. While moving ahead with new aircraft, Beech also diversified into nonaviation machinery and equipment, completing a contract for agricultural harvesters in the late 1940s. Production of the twin-engined Model 18 continued uninterrupted. The improved E18S "Super 18" model first flew on December 10, 1953, sustaining a comfortable niche at the high end of the business transport market. The Beech Model 50 Twin Bonanza, a modern twin with tricycle landing gear, made its first flight on November 15, 1949. The first of the postwar light twins in production, it first entered service as the L-23 Army liaison aircraft during the Korean War, but then succeeded as a business aircraft after military needs had been filled.

Walter Beech continued to lead a stable management team. The veterans Yankey, Gaty, and Wells continued on the board and were joined by Dwight Wallace, who also became the firm's general counsel. Olive Ann Beech, holding the titles of secretary and treasurer, remained influential in management. When Walter Beech died unexpectedly of a heart attack on November 29, 1950, at age 59, she became president and chief executive officer and managed the company with conspicuous success. Her nephew Frank Hedrick, who had joined Beech in 1940, became her second in command. Beech captured an important military contract with its T-34 Mentor primary trainer. Basically a Bonanza with a tandem glass cockpit and a conventional tail, the T-34 prototype first flew on December 2, 1948. The T-34 lost to the Fairchild T-31 in a 1949 flyoff competition, but T-31 production had been delayed due to funding shortfalls. Walter Beech then persuaded the Air Force to conduct a second competition, which the T-34 won. With Korean War priorities, the T-34 finally entered production in March 1953, was later adopted by the Navy, and was widely exported.

Another military contract, for the T-36A twin-engined crew trainer, resulted in a crisis. The largest aircraft ever designed by Beech, the T-36 was canceled on June 10, 1953, just as a prototype was ready to fly and a new factory had been completed. Overnight, half of the Beech order backlog was lost, and the employment level of 13,000 also was reduced by half.[23] But the company eventually made up the loss with increased subcontracting business. Beech became active in pilotless vehicles, or target drones, and would continue to have the largest proportion of military sales of the general aviation firms.

While no lightplane manufacturer was unaffected by the sudden drop in demand in 1947, Piper was the most severely affected of the major firms. Piper had produced 7,773 new aircraft during 1946 and was so optimistic that it opened a branch plant in Ponca City, Oklahoma. Piper suffered a setback when flooding caused extensive damage at its Lock Haven factory on May 28, 1946, but recovered quickly. Then the sudden disappearance of the market simply caught the firm by surprise. William Piper had expected a market slowdown, but not to the extent it actually occurred.

There was new product development with the powerful four-place low-wing PA-6* Skysedan and the small single-seat PA-8 Skycycle in 1946. The Skycycle, with a fuselage adapted from a military drop tank, proved somewhat unstable and faced a dubious market, but the Skysedan, designed by Jamouneau, showed promise, and Piper committed a major error by not moving it into production. Other experimental aircraft included the light twin-boomed pusher-engined Skycoupe and the PA-10 light amphibian. By 1947 Piper production consisted of the PA-11 Cub Special, successor to the veteran J-3 and carrying a base price of $2,195, and the three-seat PA-12 Super Cruiser and four-seat PA-14 Family Cruiser that extended from the prewar J-5B. Its product line, therefore, was concentrated at the "low end" of the market, most vulnerable to competition from cheap surplus aircraft, and still reflected prewar design technology. Clearly the company had stuck with the Cub design too long, producing 14,125 of the civil models, a record at the time.

After years of strong profitability, Piper lost in excess of $500,000 for the 1947 fiscal year. Dividends were suspended and debts mounted. Major production cutbacks and layoffs ensued. An alarmed board of directors, acting under pressure from Manufacturers Trust of New York and other major creditors, brought in the troubleshooter William C. Shriver, formerly with Chrysler Corporation, as general manager. Shriver was given total authority to reverse the decline and manage the company out of debt, and the move became a textbook case of the rescue of a failing company. Arriving in June 1947,

* Piper switched to the PA prefix for aircraft designations for the postwar market.

Shriver moved swiftly, completely shutting down manufacturing and closing the Ponca City branch plant. Employment, 2,607 in February, had already been reduced to under 1,000 by June, but Shriver pared it to only 157 by the end of the year. Remaining employees were forced to take pay cuts. Shriver moved into the office of William T. Piper, Sr., forcing him completely out of any involvement in daily affairs, and removed W. T. Piper, Jr. as assistant treasurer, reassigning him to sales.[24]

Shriver discovered that the conservative senior Piper had developed little in terms of modern business practices, such as cost accounting and market analysis, in addition to continuing an outdated product line. Piper's sobriquet as the Henry Ford of personal aviation proved to be accurate in less flattering ways as well. Some felt, however, that Shriver made the younger Piper a scapegoat for the financial problems in order to preserve the reputation of his father. Although removing most officers, Shriver was impressed by Walter Jamouneau and retained him as an executive and a member of the board of directors. He then pushed development of the PA-15 Vagabond, with side-by-side seating, as the best short-term prospect for sales in the diminished market.

The next task was to reassure creditors. The engine supplier Continental Motors had threatened to force the firm into involuntary bankruptcy, but Shriver persuaded Continental and others to hold off action. He also cultivated the necessary political contacts to secure a $390,000 RFC loan, enabling Piper to resume production and to use up its considerable inventory of materials.[25] Restarting production in January 1948, Shriver began expanding the workforce and succeeded in containing losses and repaying creditors. The Vagabond, while an austere and uncomfortable design, did manage to sell well enough during 1948 to generate needed cash for developing new models.

Shriver, who had announced that he intended to stay only 18 months, in one of his final acts acquired the money-losing Stinson Division from Consolidated Vultee on December 1, 1948, for 100,000 shares of new Piper stock. All Stinson assets, rights, and unsold inventory of some 200 airplanes were included. William T. Piper, Sr. opposed the Stinson deal, but Shriver felt that the acquisition would give the company broader market coverage, and the Voyager was the lowest-priced of the four-seat models on the market. Piper completed a small number of new Voyagers from assemblies, but turned down the idea of new production.[26]

With the Voyager added to Piper's PA-15 and new PA-17 Vagabond trainer, and with new four-seat developments nearing production, market coverage was indeed enhanced and recovery was underway by the end of 1948. The last Voyager was delivered in 1950, marking the end of the Stinson nameplate after 24 years and 12,320 aircraft, of which more than 5,000 were Voyagers. Shriver left at the end of 1948, having earned a reputation as an ogre, but clearly having stabilized operations. Later, the senior Piper grudgingly credited him with having saved the company.

Product changes continued as the veteran Cub Special ended production, succeeded in 1949 by the improved PA-18 Super Cub, which would enjoy a long production life. Korean War military orders for the Super Cub as the L-18 and L-21 helped the company significantly, and an agricultural spraying version also found a market. Joining the Super Cub in 1949 was the four-seat PA-16 Clipper, developed from the Vagabond. Running afoul of Pan Am's airliner trademark, the Clipper was redesignated the PA-20 Pacer and was produced in quantity from 1950 to 1954. The tricycle gear version, the PA-22 Tri-Pacer, was introduced in 1951 and became one of the most popular of all postwar light aircraft.

After Shriver's departure, William Piper still did not control the board, and his New York financiers imposed a new executive vice president and general manager, August Esenwein. More friction between the Piper family and its bankers ensued. Among other measures, Esenwein removed Walter Jamouneau as chief engineer. He also pressed for an entry into the twin-engined market, where Piper's lack of engineering expertise had led to consideration of the Baumann Brigadier. In the meantime, the senior Piper had acquired more stock in the open market, and by the board meeting of January 10, 1950, regained control and reinstated Jamouneau. Esenwein resigned.[27] Piper, although nearing 70, had no thought of retirement, but did install his second son, Thomas ("Tony"), as administrative vice president and *de facto* chief executive on April 1, 1950. Tony Piper proved to be as conservative and frugal as his father. The elder Piper went on to gain his twin-engined pilot rating in 1953 at age 73.

Piper entered the twin-engined business segment with a new design, probably the riskiest product decision yet undertaken by the company. Originally called the Twin-Stinson, possibly in tribute to an earlier Stinson design study, the prototype, with twin tails, first flew on March 2, 1952. The developed PA-23 Apache soon entered production after Piper undertook the first market survey in its history.[28] The Apache became an immediate success and led to further Piper twin-engined developments. At some $32,000, the most expensive aircraft ever built by Piper, the Apache still was priced well under the larger but competitive Beech Twin Bonanza and new Cessna 310. Max Conrad's publicized transatlantic solo flights in the Apache were a major boost to general aviation and to Piper, which became once again a strong competitor.

The General Aviation Helicopter

The helicopter had entered military service toward the end of the war, and with peace many saw an enormous civil market potential. The strong overlap in development between military and civil models aided the civil market. Manufacturers progressed with both simple, light designs and with larger passenger- and cargo-carrying models. Low-priced helicopters were forecast as the answer to the dream of the "family flivver," taking off and landing in front yards and eventually displacing fixed-wing light aircraft. Other civil uses for the helicopter included business travel, medical emergency evacuation and transportation, law enforcement, and agricultural spraying. But the dream of a light helicopter for the general public would prove to be unattainable, as had been the case earlier with the autogiro. Development efforts were highly active, however, as the following survey recounts.

The first helicopter certificated for civil operation in the world was the Bell Model 47, on March 8, 1946. While primarily a military firm, Bell Aircraft Corporation had begun helicopter development in 1942. It placed the Model 47B in civil operation in 1948 at an initial price of approximately $21,500. A total of 78 of the initial model was built, including crop spraying and cropduster versions. The New York Police Department was among the early customers. The Model 47 also was widely exported, further strengthening Bell's market position. The larger five-seat Model 42 was also planned for the civil market, but only three test models were built as the type did not reach the market.

Sikorsky, the leading helicopter manufacturer, did not target the civil market, but limited numbers of its S-51 and larger S-55 series were ordered for industrial use in the early 1950s.

The young helicopter pioneer Stanley Hiller formed Hiller Aircraft Company in California in 1942 as a division of his father's Hiller Industries. In partnership with Kaiser Industries, he tested his first XH-44 model in August 1944 with a view toward its military potential. Renamed United Helicopters, Inc., in 1945 for the postwar market, Hiller tested the experimental UH-4 Commuter civil model, which featured coaxial rotors, and the J5 model, with jet deflector vanes at the tail for control. Production success came with the two-seat Model 360, with a conventional main rotor plus tail rotor, which first flew in 1947. In 1948 it became the second helicopter to gain CAA certification. The UH-12 military variant became the standard training helicopter for both the Army and Navy during the Korean War, giving Hiller a sound production base. Then the UH-12B developed civil variant gained certification in 1951, but production waited until military needs had been met. The company was renamed Hiller Helicopters in 1950, and while primarily military, it continued civil production and development. The small Hiller HJ-1 Hornet, powered by rotortip ramjet engines, was first tested during 1951–1952 and was publicized as a potential mass-market helicopter, but it never saw production. Hiller continued to stress affordability and ease of operation, and the later Model 12C found a share of the small civil market of the mid-1950s.

Siebel Helicopters was formed in Wichita in 1948 by Charles M. Siebel, a Wichita native and former Bell engineer, specifically to pursue a small helicopter for the general aviation market. His two-seat Model S-3 showed promise, and Cessna took over the firm on March 1, 1952, in order to diversify into the market, the first established general aviation firm to do so. Charles Siebel was retained by Cessna to further develop the basic design, redesignated CH-1, which first flew in July 1954.

Newby O. Brantly founded a firm for helicopter development in Pennsylvania in 1943, and developed and tested a small coaxial design in 1946. The single-rotor production Model B-1 was the lowest-priced helicopter in the United States, leading efforts to make the helicopter more affordable. After moving production to Oklahoma, Brantly developed the Model B-2, which first flew on February 21, 1953, and led to further development of the small, affordable concept. But the company struggled to attain quantity production.

The Market Stabilizes

Despite recurring upheavals, general aviation manufacturing reflected one consistent pattern, that of entrepreneurship. The light aircraft field, however cyclical and risky, was the only remaining sector of the industry that provided entrepreneurial opportunities. Regardless of prevailing market conditions, and in the face of numerous exits by older firms, new ventures appeared. Whatever their educational backgrounds and financial backing, those forming new aircraft production ventures tended to be mechanical tinkerers and innovators, always believing that they could build and sell a better airplane. Entry barriers were formidable, though still not as high as in other sectors of the aircraft industry. Yet despite numerous and innovative new light aircraft designs, few enjoyed sustained production. Most longer-term personal aircraft successes were those of conventional configuration. The following account traces the more significant new entries.

Colonial Aircraft of Maine was organized in 1946 by David B. Thurston, a former Grumman engineer with extensive amphibian design experience, and other associates. The firm produced the C-1 Skimmer pusher-engined light amphibian, developed from the Grumman Tadpole with which Thurston had been involved at Grumman.[29] The Skimmer first flew on July 17, 1948, but did not enter production for several years.

Fletcher Aviation Corporation of California, starting in 1941 as a war subcontractor, developed light aircraft designs, although primarily for military rather than civil markets. Its first design, the wartime FBT-2 light trainer, of extremely simple configuration and of plastic-plywood composite construction, did not attain production. In the postwar period the company designed the FL-23 liaison aircraft, followed by the FD-25 Defender light ground attack aircraft in 1952, both intended for the Army. The FL-23 lost to the Cessna L-19 for orders, but the developed FD-25B was licensed by Toyo in Japan. Fletcher later turned to civil market developments.

Mooney was incorporated on June 18, 1948, in Wichita, by former Culver executives Al Mooney and Charles G. Yankey, with Yankey providing the financial backing and serving as president. Al Mooney's brother Art also was associated. Its first design was the diminutive single-seat M-18 Mite, but despite an initial price of $2,000, sales of the Mite were disappointing. Facing limited room for expansion in Wichita, Mooney moved to Kerrville, Texas, in February 1953 and reestablished operations. Mite production was phased out, but the new M-20 made its first flight on August 10, 1953. The M-20, while offering modern features and high performance, was simpler, with extensive use of plywood in its structure, and less expensive, at around $12,000, than competitive models. It would make Mooney a significant factor in the four-seat low-wing market.

An innovative entry was the Koppen-Bollinger Aircraft Corporation, organized in Massachusetts in 1948 by Professors Lynn L. Bollinger of the Harvard Business School and Otto Koppen of MIT to develop a practical short-takeoff-and-landing (STOL) utility aircraft. Both had extensive experience in aviation. Soon changing the name to Helio Aircraft Corporation, they developed a small STOL model based on a converted Piper Vagabond. Named the Helioplane, it first flew on April 8, 1949. An early proposal for Aeronca to produce the Helio design was not carried out, and it remained experimental. But a new, fully developed STOL design, the Helio Courier, was manufactured from 1954 under contract by Mid-States Manufacturing Corporation of Pittsburg, Kansas. The owners had an eye to the military liaison market as well as the civil market, and they saw limited success in both.

With the end of the war in 1945, the Defense Plant Corporation no longer required its giant factories in Dallas and Fort Worth, among others. The Fort Worth plant remained with Consolidated Vultee, but the Dallas "A" and "B" plants used by North American reverted to the RFC for disposal. The engineers Robert McCulloch and H. L. Howard, who had run the North American Dallas operations during the war, saw an opportunity to enter the light aircraft market and take advantage of the availability of the Dallas facilities. Attracting outside investors, they established the Texas Engineering and

Manufacturing Company, Limited, on November 17, 1945, using the Dallas "A" plant. Aeronautical terminology was deliberately left out of the name for later diversification purposes.

The company began production with the Fairchild F-24 and the Globe Swift, having acquired the rights to both designs. In late 1946 the firm was incorporated as TEMCO, Inc. Globe Aircraft Corporation had been optimistic about the postwar market but soon encountered financial difficulties. The Globe bankruptcy was an early crisis for TEMCO, but it acquired Globe's assets and inventory on June 23, 1947, and continued production of the Swift, a classic light aircraft design, from 1947 to 1950.

TEMCO went on to acquire the assets of the bankrupt Luscombe Airplane Corporation of nearby Garland, Texas, in 1949. The reorganized Luscombe Holding Company became a wholly owned subsidiary on January 12, 1950. TEMCO resumed Silvaire production in March 1950, producing 50 before again suspending it for military work. Then all light aircraft production was suspended in October as capacity was needed for Korean War subcontracting requirements, and personal aircraft were officially discontinued, effective December 31, 1950.[30]

TEMCO soon began development of original designs, including the T-35 Buckaroo trainer developed from the Swift, and again was reorganized as the Temco Aircraft Corporation in 1952. Luscombe was fully merged into Temco on April 2, 1953. Additional factory space was leased at Greenville, Texas. Production success did not follow for Temco, since the T-35 lost out to the Beech T-34, and only 20 were ordered. But Temco did achieve success with adaptations. The entrepreneur Jack Riley completed the first Navion twin-engined conversion in April 1952, naming it the Twin-Navion. Temco purchased the Riley Twin-Navion operation in 1953 and further developed the conversion as the Riley Twin. Temco produced substantial numbers of the conversions from that year. A separate line of conversions was undertaken from 1953 by Cameron Iron Works of Galveston, Texas, as the Camair 480, with more powerful engines than the Temco variant. The company established the Camair Division for the operation.

In the fall of 1944, while the nation was still at war, the Douglas engineer Theodore Raymond (Ted) Smith, an industry veteran and a major contributor to the design of the A-20 Havoc, already was thinking of the postwar civil aviation market. With a small number of Douglas associates, Smith organized the Aero Design and Manufacturing Company in Culver City, California, on December 21, 1944. His objective was to develop a completely new twin-engined light executive transport, to be called the Commander, that would be an advance over anything developed to that time. Smith and several of his colleagues resigned from Douglas on August 15, 1945, to pursue the venture full-time but soon experienced problems obtaining financing. Initial capital raised was only $175,000. Consequently, the first prototype, designated L3805 (for Lycoming, 380 horsepower, five-place), did not fly until April 23, 1948. It showed much promise and several orders were received, but again production was held up due to financing shortfalls.

The company survived by attracting financing from the Pew oil interests of Philadelphia and the Amis brothers construction interests of Oklahoma City. With civic support from Oklahoma City, the company was reorganized in September 1950 and relocated to a new factory there.[31] Ted Smith held the position of vice president, research and development. The Aero Commander 520, the developed production version, first flew on August 25, 1951. The first delivery to a customer was on February 5, 1952, and 46 were delivered by the end of the year, the beginning of a more than 30-year production life for the Commander series.

# # # # #

During 1949 there was significant price cutting by most manufacturers in the small industry to maintain demand. But by 1950 the general aviation market could be said to have stabilized, although at a production level much lower than that of even the most pessimistic forecasts of 1945. Such respected names as Aeronca, Ercoupe, Luscombe, Stinson, and Waco had exited. In 1951 only 2,302 personal airplanes were produced, and no more than five significant manufacturers, Beech, Cessna, Piper, and the new Mooney and Aero Design firms, would survive through the decade.

Private flying no longer gripped the popular imagination by the 1950s as it had between the wars, and private aircraft ownership simply was not a widely held aspiration.[32] Yet the Civil Aeronautics

Administration still estimated that the active civil aircraft fleet had grown to some 54,000 by 1953, of which only about 1,500 were commercial airliners.[33] CAA Administrator Theodore P. Wright undertook special efforts to promote private flying from 1952. But the CAA, also responsible for the certification process for new aircraft, lacked the resources to perform all tests during the late 1940s and delegated much testing to the manufacturers themselves. The CAA then relied on the test results, but that practice would later engender safety controversies.

The Aircraft Industries Association introduced the utility category and formed the Utility Airplane Council in 1952, in succession to the Private Aircraft Council. Dwane Wallace served as the first chairperson. The AIA formally identified six utility roles: business flying, industrial (including pipeline patrol, aerial mapping and survey, forestry, etc.), agricultural, instructional, charter, and pleasure or sport. Noteworthy was that the last category, once the only general aviation role, had become the smallest. The AIA ended use of the term "personal planes," in part to change the public's perception of private flying as the province of the country club set or of wealthy playboys, but also in recognition of the dominance of business flying.

The business market segment began to grow rapidly, and new twin-engined executive aircraft such as the Aero Commander 520, Twin Bonanza, Piper Apache, and Cessna 310 helped expand the market. In addition, the export market for general aviation aircraft became a significant factor for the first time. Beech established an export division as early as 1946. Light aircraft were exported to numerous countries, and the export component of total sales rose rapidly. The American general aviation industry remained the *de facto* world general aviation industry. The British industry, for example, very active in personal aircraft during the 1930s, did almost nothing in the field postwar.

The cost of general aviation aircraft would continue to rise, although limited by market pressures. There was little doubt that price was very much the decisive factor in personal and business aircraft marketing. For larger commercial aircraft, delivery dates, operating characteristics, and financing arrangements could be more influential, but for the general aviation customer, price remained paramount.[34]

5

General Aviation Matures, 1954–1967

The year 1955 was a boom year for all civil aviation. Business flying increasingly became regarded as the future of general aviation, and marketing efforts focused on business applications (see Table 5-1). The Corporate Aircraft Owners Association, founded in 1947, became the National Business Aircraft Association (NBAA) in 1953 and advanced the field. Twin-engined aircraft became more prominent as larger, luxurious types became increasingly preferred for

Table 5-1
Annual Production Sales, General Aviation Aircraft

Year	Units	Factory sales ($ millions)
1955	4,434	$ 68.2
1956	6,738	103.7
1957	6,118	99.6
1958	6,414	101.9
1959	7,689	129.8
1960	7,588	151.2
1961	6,788	124.2
1962	6,697	136.8
1963	7,569	153.4
1964	9,336	198.8
1965	11,852	318.2
1966	15,768	444.9
1967	13,577	359.6

NOTE: Figures account for fixed-wing aircraft only.
SOURCE: *General Aviation Statistical Databook* (1990–1991 edition), p. 4.

business use. While smaller single-engined aircraft still dominated numerically, sales of light twin-engined models rose from 354 in 1954 to 870 in 1958.[1]

Light aircraft design advanced but almost entirely by industry-funded research. Major NACA effort and focus from the 1950s was on high-speed research, irrelevant to general aviation until the advent of business jets. Yet general aviation still reflected basic technology developed by NACA from the 1920s. However limited, most NACA general aviation research was undertaken by the Langley center in Virginia.

Beech, then the largest general aviation firm, reported sales of $76 million in 1955, followed by Cessna with $50 million and Piper with $17 million. Yet general aviation remained a minor sector of the broad aircraft industry, as such smaller military firms as Republic, Martin, and Grumman each reported sales in the $300 million range for that year.

Light aircraft production expanded steadily in the prosperous economy of the 1950s and 1960s. The term *general aviation,* referring to all civil aviation outside the scheduled commercial transportation sector, came into popular use around 1957–1958. To that point the term *utility aircraft* had been preferred, although such terms as *private aircraft, personal aircraft,* and especially *business aircraft* were still widely used. The public began to recognize the practical value of general aviation.

The change coincided with the redefinition and expansion of the overall aircraft industry into the aerospace industry, as missile and space programs became more important. Smaller commuter aircraft and commuter airline operations also began to be included in the general aviation sector. But a point of possible confusion remained in that helicopters were regarded as a distinct sector of the aerospace industry, although the role of the helicopter in general aviation was increasing steadily. General aviation helicopters continued to be treated separately from fixed-wing aircraft in statistics.

General aviation manufacturers continued to be strengthened by military demand for training and utility aircraft. Those aircraft often overlapped strongly with general aviation designs and could be purchased "off-the-shelf" at considerable savings in research and development expense. Thus general aviation firms could easily and advantageously fill military needs and potentially reap an added benefit from a countercyclical effect. Military demand remained particularly vulnerable to budgetary constraints, however, as training and utility aircraft generally held a lower priority than combat and other larger types. The Beech T-36 cancellation was a case in point.

Pleasure and recreational flying remained too expensive for development of a mass market. All active producers struggled to gain and hold market share, but competition was so intense that most smaller firms with less complete product lines could not survive against the major producers. As a result, general aviation manufacturers declined steadily in number, some by failure and others through merger. But newer competitors persevered, feeling that a new or innovative design could gain a profitable market niche.

The Homebuilt Segment Emerges

One response by sport or recreational pilots to the expense of private flying and to the decline of sport or recreational flying was that of constructing, and in some cases designing, their own airplanes for personal use. Although not new, the business of supplying kits of components or plans for amateur builders began to gain some momentum in the postwar environment. The market niche was potentially aided by the lack of interest in the field by the established factory aircraft producers.

Government regulatory authority had been largely negative on the matter of homebuilt or amateur-built aircraft, even for personal flying. The Bureau of Air Commerce and later the CAA had simply made no provision for certification of homebuilts, which consequently could not be granted registration numbers. An experimental, or "X" registration number, primarily intended for established manufacturers to test a new type, was valid for only 30 days.

The CAA gradually began to become more supportive of amateur aircraft after the Second World War and also encouraged personal aviation. A leading amateur aircraft designer and builder, George Bogardus, lobbied the CAA during 1946 and 1947 for rule changes which would facilitate and encourage homebuilts. With the change in the regulatory environment since the 1930s, Bogardus was successful. The CAA established a permanent category for homebuilts in 1949. The ruling permitted X registrations for six months' operation of airplanes that had not been certified by the CAA, and the certificate was renewable. The certification applied to aircraft that were not used for commercial purposes and to those that were at least 51 percent built by the owner-operator. In addition, they could not be built for resale. The CAA regarded such aircraft as falling within the experimental category, and that proved a major incentive to development of a kitplane or homebuilt segment. After being tested according to specified standards, a homebuilt aircraft could be awarded a type certificate by the CAA Aviation Safety Agent in the field.

More specific Civil Air Regulations (CARs) regarding homebuilts became effective January 15, 1951. Then a manual with procedures for issuance of airworthiness certificates was officially adopted on September 19, 1952. The legal basis for development of amateur-built or homebuilt aircraft was complete.[2] The change was also consistent with overall CAA efforts to promote private flying, and was especially important given the cost of recreational flying. The change spurred more widespread amateur experimentation in aircraft, much involving aircraft of original design.

Sport or recreational flying with aircraft of amateur construction remained the smallest segment of general aviation, but regulatory progress and the more supportive attitude by the CAA encouraged the field. A sufficient core of enthusiasts had developed that the Experimental Aircraft Association (EAA) was founded on January 26, 1953, in Milwaukee, Wisconsin, by Paul H. Poberezny, a combat veteran pilot and leading experimenter. Under Poberezny's leadership, the EAA became the umbrella organization for the field and grew steadily, expanding internationally and eventually claiming more than 160,000 members worldwide. The term *experimental,* while perhaps misleading, was defined by the CAA generally to include aircraft of original design and construction for personal use rather than for business use or commercial sale. But experimental aircraft had entered the general aviation vocabulary.

The year 1953, which marked the founding of the EAA, also could be regarded as the modern founding date of the homebuilt segment of general aviation. The segment, always characterized by experimentation and entrepreneurship, had seen numerous efforts toward development of a distinct homebuilt or kit airplane business during the period from the end of the Second World War to 1953. Legal and market barriers to entry into the field were low, and required investment also was comparatively low. Further, those firms involved with factory-built general aviation aircraft continued to ignore the kit or homebuilt field.

Among the pioneers in small homebuilt aircraft was Ray Stits, who began experimenting in 1948. He founded Stits Aircraft in Riverside, California, and developed his Playboy and Flut-R-Bug models. Stits also claimed the distinction of designing and building the "world's smallest airplane." He sold both plans and component kits. The Playboy entered the market in 1953 but experienced weak demand, a situation faced by all small single-seat sport models, regardless of merit. Mooney had had a similar experience with its factory-built Mite. The larger mid-winged Stits Flut-R-Bug recorded sales, less engine/propeller assemblies, of 27 kits by 1956.

Another pioneer was Curtis Pitts. He designed his single-seat S-1 Special during 1943-44 and first flew it in September 1944. Pitts built a series of the design over the years. Plans were available for home construction, but the Special did not become available in kit form until the 1980s.

A third notable pioneer was Sylvester "Steve" Wittman. The Wittman W-8 and W-10 Tailwind, small two-seat cabin monoplanes developed during 1952-53, developed a strong reputation.

A fourth entrant, the Ace Aircraft Company of West Bend, Wisconsin, was the successor to the early firm of O. G. Corben. Ace modernized the original Corben Baby Ace of 1931, a single-seat model with a high parasol-type wooden wing adaptable for home construction. Corben assets were acquired by Paul Poberezny in 1954, and his redesigned Baby Ace, using some Cub components, first flew on November 15, 1956. Over 350 kits were sold and completed. In 1961 Poberezny sold the Ace firm to Edwin T. Jacob, who continued marketing the design. In 1955 the EAA also designed and marketed its own homebuilt design. EAA efforts continued into the 1970s.

Another popular early design for home construction was that of Robert Nesmith of Houston, Texas. The Nesmith Cougar, a conventional high-wing tailwheel design, first flew in March 1957, and several thousand sets of plans were sold.

Throughout the 1950s other amateur builders constructed numerous small aircraft, some from original designs, but most as experimental models rather than as a basis for sale as plans or kits. A popular alternative which has continued was to copy designs of classic light aircraft of the 1920s, 1930s, and 1940s, or to develop smaller replicas of wartime aircraft such as the P-51 Mustang. The biplane configuration remained especially popular for single-seat sport models.

Despite continuous activity and new entrepreneurial efforts, development of the homebuilt or kit segment progressed very slowly through the 1950s and 1960s. The major incentive remained that private aircraft ownership was too expensive for most pilots, with kit aircraft offering major cost savings over new factory-built aircraft, even of such small models as the Super Cub. Also, many amateur builders took particular pride in constructing their own airplanes. Power by modified Volkswagen air-cooled engines was popular in amateur aircraft.

Yet obstacles and disincentives were many. Home construction could encounter delays and technical difficulties. The completion rate from kits was low, and the rate from plans even lower. Aviation periodicals frequently ran advertisements offering partially completed kit aircraft for sale by those who had abandoned the attempt. In addition to construction problems, the economics of the field remained unfavorable in that good used factory-built aircraft still could be cheaper than home construction when all expenses were factored in. That availability proved a major deterrent to homebuilt market growth. In addition, there was a general lack of an aftermarket for homebuilts, especially single-seaters, whatever their quality. And despite the apparent appeal of a single-seat sportplane, a significant commercial market still had not developed.

Technical factors and safety issues presented further problems to growth. The certification process for homebuilts remained a major hurdle. Homebuilts, being small and light, were very sensitive in the weight-and-balance area, with little margin for error. Further, the accident rate was very high during the initial test period; afterward the safety record improved markedly. Many accidents were attributed to construction shortcuts or to unapproved modifications, but there was a troubling incidence of structural failure. These factors combined to limit the development and growth of the homebuilt field until the 1980s.

The economic impact of the homebuilt or kit construction segment during the 1950s and 1960s was difficult to measure. No organization or agency maintained complete statistics on the field; national sales figures for plans, kits, or for the value of completed aircraft were unattainable. Yet there was little doubt that the importance of the segment, however obscure and however gradual, was increasing. Illustrating later growth, *Kitplanes* magazine reported that 25,401 kits from manufacturers were sold in the United States from the years 1963 through 1989.[3]

Industry Survey

With business aviation expanding and aircraft demand growing in a generally prosperous economy, established firms as well as newer competitors benefited. Production rose steadily and new models appeared.

Mooney, with a promising four-seat retractable-gear design in the Mark 20, faced the problem of financing its production. Charles Yankey died, and Piper, seeking a more modern four-seat design for its line, considered purchasing the rights to the Mark 20, but negotiations faltered. Al Mooney later expressed regret at not coming to an agreement with Piper when he had the opportunity.[4] Al W. and Art B. Mooney then were forced to sell all stock to outside investors in order to save the design. Under new ownership, Hal F. Rachal became president on July 1, 1954, and the Mooney brothers joined Lockheed as engineers in 1955.

The restructured firm restored the Mite to production for a time, and the Mark 20 was certified in August 1955. The Mark 20 gained popularity and was joined by the Master, a fixed-gear version, and then by the advanced Mark 22 Mustang, with retractable gear and pressurization. The Mark 22 was a market failure, however; only 30 sold before production was suspended. But Mooney steadily built a niche in the low-wing, four-place market, with the simpler, lower-cost Mark 21 joining the product line in 1961. Mooney enjoyed a record production year in 1965, and in 1966 began distributing the Mitsubishi MU-2, a new high-winged twin-engined utility and executive aircraft which had first flown on September 14, 1963. Mitsubishi had established a production subsidiary in San Angelo, Texas, and not far from Kerrville, for the U.S. market. The MU-2 attracted steady orders.

Aero Design produced 150 of the initial Model 520 by 1954 before switching over to the improved Models 560 and 560A. The company suffered a disastrous fire in August 1957, but with community support resumed production in four months.[5] More advanced developments followed in succession, including the 680 Super and the pressurized 720 Alti-Cruiser of 1958, although only 13 Alti-Cruisers were built. With higher performance and increased size, these models grew further away from the original concept, and Aero Design accordingly developed the smaller Aero Commander 500 and 500A models, named Shrike Commander, for the "low end" of the twin-engined segment. Three liaison models were supplied to the Army as the YL-26A, while the Air Force acquired 15 L-26Bs as VIP transports. President Eisenhower's publicized use of an Aero Commander, alleviating lingering safety concerns, was a further boost to business flying.

Aero Design was sold to Rockwell-Standard Corporation of Pittsburgh, principally in automotive parts and other machinery, in June 1958, as that firm entered aircraft manufacturing as part of a diversification strategy.[6] The subsidiary name was changed to Aero Commander, Inc., in October 1960, and sales success in the twin-engined segment continued. Current models were refined and new developments proceeded. The lengthened Grand Commander 680F/L, which first flew on December 29, 1962, marked a major advance into the luxury executive market. A further development was the Turbo Commander, powered by twin Garrett AiResearch TPE 331 turboprops, which first flew on December 31, 1964.

A strong trend of reviving older light aircraft designs for further development and production continued. Such revivals often were under new corporate auspices where the original firm had entered bankruptcy or otherwise exited the field. Rights to older designs also were sold to new or existing producers. CallAir, for example, produced the Super Cadet, developed from the wartime Interstate S-1A Cadet, while continuing limited production of its A-2 and A-3. CallAir began agricultural aircraft development in 1953, one of the first firms to offer a specialized model for this growing segment. The Model A-4 first flew on December 14, 1954. The Models A-5 and A-6 were agricultural adaptations of the A-4 and achieved some success from 1957 in that role, to that point performed by conversions of such types as the 1930s Stearman trainer.

On January 1, 1959, the company was incorporated as CallAir, Inc., with capital of $1 million. The Call family then sold the corporation to John Mangum, a former distributor, on November 11, 1959. Unfortunately, the arrangement did not endure as business, always low-volume, declined further and dragged the firm into receivership. Two new investors purchased the assets for only $25,000, which they in turn sold to the Intermountain Manufacturing Company (IMCO) at a public sale in February 1962. IMCO then resumed production. IMCO continued to develop the agricultural aircraft series, leading to the CallAir A-9, which enjoyed considerable sales success during 1963–1965.[7] CallAir rights and assets were acquired from IMCO by Rockwell-Standard in December 1966. To amalgamate its growing small aircraft activities, Rockwell-Standard closed the Afton, Wyoming, plant in September 1967 and transferred remaining production to a new factory in Albany, Georgia.

Northern Aircraft, Inc., which had purchased design rights for the Model 14-19 Cruisemaster from Bellanca in 1955, entered production with the 14-19-2 in 1957. The firm changed its name to Downer Aircraft Industries, Inc., on January 1, 1959, and continued to market and develop the basic Cruisemaster design. The Cruisemaster countered the prevalent all-metal construction of its main competitors, still using wood extensively in its structure. Production was halted on December 1, 1960, but later resumed with the Downer Bellanca Model 260A with tricycle gear. Downer in 1964 sold the Bellanca design rights to its successor, International Aircraft Manufacturing, Inc., known as Inter-Air. Inter-Air continued to produce at the same location a further development of the Model 14-19 as the Miller Bellanca 260-A, Miller being the distributor. In 1966 another reorganization resulted in Inter-Air becoming the Bellanca Sales subsidiary of Miller Flying Service.

In 1956, after the sale of Model 14-19 rights, Giuseppe M. Bellanca and his son August T. Bellanca formed a new Bellanca Aircraft

Engineering, Inc., at its old Delaware location. The firm held rights to all designs other than the Model 14-19 and pursued new developments. Giuseppe Bellanca died in 1960, but August Bellanca continued advanced light aircraft development.

Champion Aircraft Corporation, a subsidiary of Flyers Service, Inc. a fixed-base operator, was established in June 1954 in Osceola, Wisconsin. Having acquired rights from Aeronca for the Champion, the company resumed production of the Champion 7 in February 1955. Improved as the Traveller, it was produced in some numbers and was joined by the tricycle-geared Tri-Traveller. Both were produced until 1964, when production switched over to the further modernized Citabria (airbatic spelled backwards). In 1961 Champion developed a small twin-engined, two-seat model, the Lancer 402, the lowest-priced twin on the market, but it was not produced in quantity.

Rights for the Ercoupe, which had ended production in 1950, were sold to Fornaire Aircraft Company of Fort Collins, Colorado, in 1955. Fornaire was an affiliate of Forney Manufacturing Company, a farm machinery maker dating from 1932. Fornaire first tested its improved F-1 Aircoupe on February 18, 1956, and later offered three different models of the basic design, but volume sales still failed to develop. Fornaire rights were sold to the city of Carlsbad, New Mexico, in August 1960, which leased the rights to the Air Products Company. The first new Aircoupe flew on December 16, 1960, but the venture ended in 1962. Fornaire and Air Products combined sold only some 30 Aircoupes. The City of Carlsbad then sold design rights to Alon, Inc., of Kansas, organized on December 31, 1963. Alon was a venture of former Beech engineers John F. Allen Jr. and Lee O. Higdon, formed to produce the successor Air Coupe. While considerably modernized, the Air Coupe still could not compete with the Cessna 150 in the small two-seat segment. The four-seat A-4 development first flew on February 25, 1966, but was not produced.

Culver rights and assets were sold to Superior Aircraft Company, which had been formed in mid-1956 as a division of the Priestley Hunt Aircraft Corporation in Culver City, California, to continue the line. The Culver Model V, dormant for 10 years, was developed as the Superior Satellite. The first Satellite flew on December 20, 1957, but only some 40 were produced.

A new Silvaire Aircraft Company of Fort Collins, Colorado, was formed by entrepreneur Otis T. Massey in January 1955, with the purchase of Silvaire rights from Temco. Production of an improved model was undertaken and sold in small numbers from 1956 to 1960, but the Silvaire company exited in 1962. Don Luscombe, out of the industry for many years, died on January 11, 1965, aged 69.

Clayton Brukner sold Waco, inactive in aircraft production since the Second World War, to Allied Aero Industries in 1963. The Troy, Ohio, plant was closed soon thereafter, and the Waco nameplate disappeared. A new Navion Aircraft Company was formed in Galveston, Texas, to produce the modernized five-seat Rangemaster version of the original Navion. The Rangemaster first flew on June 10, 1960, and entered small-scale production. In mid-1965 this venture was succeeded by a new Navion Aircraft Corporation of Seguin, Texas, organized by the American Navion Society, an association of Navion owners. The company produced a more powerful Rangemaster but also encountered market difficulties.

From the mid-1950s the Big Three, Beech, Cessna, and Piper, steadily introduced new, higher-performance designs on the market, spurred by overall market expansion and their own financial strength. Each made significant aeronautical advances, and Cessna and Beech also developed strong military aircraft and subcontracting businesses. The general aviation sector, especially the Big Three, developed product line proliferation during this period, even instituting annual model changes in the manner of the auto industry. By 1967 Cessna, for example, was producing no fewer than 28 commercial models in addition to military types.[8] Many were of the same basic design but differed in engine power, interior appointments, or equipment. The civil aircraft inventory exceeded 100,000 for the first time in 1966, and continued growth prospects and attempts to meet all customer requirements and serve all segments of the market encouraged such developments.

Piper continued production of the Tri-Pacer and Super Cub throughout the 1950s, and the success of the Tri-Pacer led the company to strong profits. A total of 7,668 of the Tri-Pacer series was produced, making it one of the most popular of all light aircraft. Piper remained a distant third behind Beech and Cessna in size, however. Piper produced the two-seat PA-22 Colt, similar to the Tri-Pacer, from 1960, and followed its successful twin-engined Apache with the similar but slightly larger PA-23-250 Aztec in 1959, and with the refined PA-23-235 Aztec in September 1962. The Apache, along with the later Comanche and Cherokee, marked the beginning of the Piper tradition of Indian names for its aircraft, although this led to some confusion with the U.S. Army, which followed the same practice in helicopter names. In 1960 Piper gained a military order for the first time in several years, with 20 Aztecs supplied to the Navy as the U-11A.

In considering an eventual successor to the Tri-Pacer, Piper decided on a low-wing design. It was the firm's lack of extensive research and development resources which had led earlier to consideration of the Mooney M-20, but unsuccessful negotiations led Piper to the original design project in 1955. Also at this time Piper began to look for an expansion site and favored Florida for its labor supply, weather, and union-free environment, a particular requirement of the elder Piper. The Piper Engineering Center was opened at Vero Beach, Florida, in 1957, and production followed. The PA-24 Comanche, with low wing and retractable gear, first flew on May 24, 1956. Later the PA-28 Cherokee, similar to the Comanche with fixed gear, entered production, succeeding both the Tri-Pacer, which ended production in 1961, and the smaller Colt, produced until 1964.

The Cherokee, in both two-seat and four-seat versions, was the first product of the new Vero Beach factory, and it became a best-seller. The PA-30 Twin Comanche followed in 1961, eventually replacing the Apache. Piper developed a specialized agricultural aircraft in the PA-25 Pawnee, which entered production in 1959 and became a strong competitor in that growth segment. By 1964 Piper production was entirely low-wing except for the veteran Super Cub. Production capacity expanded both in Pennsylvania and Florida, and employment reached a record 4,000 during 1966.

Fred E. Weick, designer of the Ercoupe and a pioneer of the agricultural aircraft, joined Piper in 1957 and led agricultural aircraft development and helped design the Comanche. He remained with Piper until his retirement in 1969. Tony Piper decided to step down from the post of chief executive in June 1960, feeling that his conservatism had held back the company.[9] Bill Piper, Jr., regarded as a more people-oriented manager, succeeded him as executive vice president and chief executive, while the senior Piper remained active as chairman of the board and president. Tony and Howard "Pug" Piper continued as vice presidents, and Walter Jamouneau as chief engineer. Even as the firm enjoyed growth and prosperity, Bill Piper, Jr. was determined to broaden the product line to meet the challenge of Beech and Cessna.[10] Among new models was the six-seat PA-32 Cherokee Six, with lengthened fuselage. Piper also entered the growing executive aircraft market with the pressurized twin-engined PA-31 Navajo. Piper's largest aircraft to date, with capacity of six to eight passengers, the Navajo first flew on September 30, 1964.

Cessna continued to broaden and improve its product line during the 1950s. Clyde Cessna, inactive in the industry since 1936, died on November 11, 1954, at age 75. In 1957 Cessna established separate Commercial Aircraft and Military Aircraft Divisions, reflecting its major activities. Cessna had lagged Beech and Piper by several years in offering tricycle-geared single-engined models, but introduced that feature in 1956 when the Model 172 replaced the tailwheel Model 170. The Model 172 was joined later by the deluxe Skyhawk version, and a further refinement was the Model 175 Skylark. The tailwheel Model 180 continued after being joined by the Model 182 with tricycle gear, and then by the deluxe Skylane. The more powerful Model 185 Skywagon utility version, still with taildragger gear, was ordered by the Air Force as the U-17A for delivery to allied nations. The Cessna 172/175 and 180/182 series soon became the best-selling personal aircraft in history, with total production exceeding even that of the J-3 Cub.

Cessna's largest business aircraft, the Model 620, configured as a small airliner with a capacity of up to 18 passengers and powered by four piston engines, first flew on August 11, 1956. The Model 620 did not enter production, however, because surplus twin-engined local-service airliners available at lower prices effectively killed market prospects. Coincidentally, North American Aviation in 1953 had considered reentering the civil market with a similar small four-engined executive model but did not proceed. In January 1957 Cessna introduced the Model 210 Centurion, a high-wing model featuring fully retractable landing gear. Then in September 1957 Cessna reentered the two-seat trainer market with the Model 150, which succeeded the old Models 120/140 and sold in large numbers. Cessna reached the milestone of 50,000 aircraft produced on February 25, 1963.

In the military sector, the L-19 (later O-1) was reinstated in production in 1961 for both the Army and Marine Corps and was also produced under license in Japan. Production in the United States totaled 3,381 for all versions by 1963. While continuing production of its successful T-37 jet trainer, which entered service in 1957, Cessna developed the YAT-37D armed light attack or counterinsurgency version, which first flew on October 22, 1963. The Air Force acquired 80 L-27A liaison versions of the Model 310. Later redesignated U-3A, military Model 310 deliveries ultimately reached 196. The two-seat Model 172E became the T-41A primary trainer, initially delivered to the Air Force in 1964 and later reordered in quantity by both the Air Force and Army (as the T-41B), and for export.

In addition to strong general aviation and military aircraft production, Cessna expanded internationally by purchasing a 49 percent share in the French firm of Avions Max Holste on February 16, 1960. The firm was renamed Reims Aviation on January 30, 1962. The Model 150 and Model 172 were produced under license by Reims. Cessna also held a major position in the metal propeller market with McCauley, which it acquired on August 1, 1960.[11]

Cessna continued to expand its market coverage in the 1960s, as Dwane Wallace determined that the company would offer an airplane to meet every need in general aviation. Model 180 developments included the Model 205A and Model 206 Super Skywagon utility models of 1962, with fixed tricycle gear, and the Super Skylane, extended to carry six passengers. A further development of the Model 310 was the Model 320 Skyknight of 1961, with turbocharged engines.

The innovative Model 336 Skymaster, with push-pull twin engines and a high-winged twin-boom configuration, first flew on February 28, 1961. Initial sales were sluggish, however, and it was succeeded in 1965 by the considerably redesigned Model 337, with retractable gear. While conceptually sound, avoiding the asymmetrical problems of one-engine operation by conventional twins, the Model 337 was rather noisy and expensive, and sales never reached the levels anticipated.

Cessna developed the Model 188 AgWagon in 1965, a specialized agricultural aircraft which competed with those of Piper, Grumman, and CallAir. Rounding out its product line was the completely new 400 series of larger business twins. The first entry, the Model 411, with six-to-eight-seat capacity, first flew on July 18, 1962, and entered service in 1965. The Models 401, 402, 411-A, and 421 were further developments of the basic design. In 1964, after 28 years as president, Dwane Wallace moved up to chairman of the board while remaining chief executive. Cessna overtook Piper in total unit production on June 1, 1967, 73,900 to 73,250.[12] Piper had built a strong lead earlier with the Cub, while Cessna did not attain high volume until after 1948. Cessna employment exceeded 12,000 for 1967.

Beech, while maintaining a commanding position in the high end of the general aviation market, still sought to broaden its market coverage. But one disappointment in the military sector was the Jet Mentor primary trainer, which Beech somewhat secretly developed with its own funds. Based on the T-34 airframe, converted to house a small jet engine, the Jet Mentor first flew on December 18, 1955.[13] Beech hoped to gain an Air Force order, but the order never materialized. With the continuing expansion of general aviation, Beech became the object of merger rumors during 1956 that involved the defense contractor General Dynamics, which had acquired Convair in 1953. Both companies denied the rumors, and in fact no merger occurred.

Beech continued production of the successful Bonanza, Twin Bonanza, and veteran E18, and introduced the Model 95 Travel Air (originally Badger), essentially a scaled-down Twin Bonanza with a Bonanza cabin, which first flew on August 6, 1956. It was joined by the Baron, a more powerful version, which first flew on February 29, 1960. Beech also produced the U-8F Seminole for the Army, similar to the Twin Bonanza but with a more capacious fuselage. The civil version of the U-8F, the Model 80 Queen Air, followed on August 28, 1958. The Queen Air was ordered by the Army as the L-23F, retaining the basic designation of the older Twin Bonanza.

The Queen Air led to a Beech bid for permanent leadership in larger twin-engined business aircraft when the company developed the airframe into the lengthened, pressurized Model 90 King Air, with twin PT6A turboprop power. It first flew on January 20, 1964.

Although many expressed concern about the high cost of turbine power, the King Air gained a market, and further developments dominated the larger executive aircraft segment. Production of the Twin Bonanza was phased out during 1964, but in that same year the ultimate "Twin Beech" development, the H18 Super-Liner, was introduced in both tailwheel and tricycle-gear versions.

For the low end of the market, Beech offered the small four-seat Musketeer, which first flew on October 23, 1961. The Musketeer III, introduced in 1965, succeeded the initial models. The light Travel Air and Baron twins continued, and an Army trainer version of the Baron was the T-42A. The Model 99 executive aircraft, a long-fuselage Queen Air development with twin turbine power, first flew in July 1966 and would lead to a long series of developments. Commercial sales for the company tripled in eight years, from $32 million in 1958 to $100 million in 1966.[14]

Olive Ann Beech began sharing more management duties with her nephew Frank Hedrick, who was promoted to executive vice president in 1960. Mrs. Beech held the office of president and retained decision authority, while Hedrick assumed a more public role. Veteran executive John P. Gaty retired for health reasons in 1960, then died in 1963. Beech operated branch plants in Liberal and Salina, Kansas, and in Boulder, Colorado, to handle both its commercial and military subcontracting businesses. As defense business gained in importance, Beech began development of supersonic target drones in 1955.

Grumman, out of civil aviation for almost a decade, decided to reenter, feeling an urgent need to diversify away from its heavy dependency on the U.S. Navy as well as sensing growth opportunities. Its successful entry into the agricultural market was the G-164 Ag-Cat, a biplane design of dated appearance but incorporating the latest agricultural aircraft technology. It first flew on May 27, 1957. Given its high overhead as a military contractor, however, Grumman licensed production to Schweizer Aircraft, heretofore exclusively a sailplane producer.[15]

The Colonial C-1 Skimmer amphibian finally entered production in 1955, enjoying an effective monopoly in its narrow segment. Colonial sold rights to a new Lake Aircraft Corporation in October 1959, which continued production as the Lake LA-4. Lake, in turn, was acquired by Consolidated Aeronautics of Indiana in 1962. David B. Thurston remained active both as a designer and consulting engineer.

Helio acquired its Kansas contractor in July 1956 and produced the more powerful Super-Courier from 1958, which was ordered as the U-10A for the Air Force and Army. The U-10A served with success in Vietnam, where its STOL capabilities were especially valuable. The Helio U-5A, a small twin-engined development, was also built in small numbers. Military orders kept the company alive as it struggled to develop broad civil demand for its specialized designs. The Courier II, a lighter, higher-performance development of the Courier I, entered service in 1965. The most advanced development, appearing in 1964, was the Helio Stallion, which was turboprop-powered and could carry up to 10 passengers.

Taylorcraft developed a new Zephyr 400, with a moulded fiberglass fuselage and wing coverings in 1955, but despite its advantages, it was not successful in the market. Fiberglass floatplane and agricultural models also were developed. The team of B. J. Mauro as president, C. Gilbert Taylor as vice president, and Jack Gilberti as chief engineer continued to manage the firm, but declining business forced a production shutdown in 1959, followed by bankruptcy. Remaining Taylorcraft assets were sold in 1963. Gilbert Taylor went to California as chief engineer for the small Saturn Aircraft venture, which was developing the former Monocoupe Meteor high-speed business twin as the Saturn Meteor II. The design did not attain production, however.

Meyers Aircraft Company of Tecumseh, Michigan, developed the more advanced four-seat low-wing Model 200 to succeed the MAC 145. The Model 200 first flew on September 8, 1953, but did not enter production until 1959. Meyers also resumed production of the MAC 145T tricycle-geared model in 1961. After the demise of Taylorcraft, Jack Gilberti launched a new venture, Volaircraft, Inc., in Aliquippa, Pennsylvania, to build a new lightweight all-metal high-wing design, the Model 10. The developed Volaire Model 1050 went into production in 1963 and gained market share.

Volaircraft, simultaneously with Meyers, became another acquisition of Rockwell-Standard, through Aero Commander, on July 12, 1965. Rockwell-Standard continued its aggressive strategy of heavy involvement in the general aviation aircraft market. Volaircraft and Meyers were purchased for approximately $1 million cash each. The Volaire Model 1050 filled the gap at the lower end of the Aero Commander range. Under the Rockwell-Standard Aero Commander Division, the Meyers Models 200/200A evolved into the improved Model 200B, which became the Aero Commander 200. The Volaire 1050 became first the Aero Commander 100, and in 1968 the Darter Commander.

Aero Commander headquarters remained in Pittsburgh, but organizational changes ensued with Rockwell-Standard's entry into the light single-engined market. Operating divisions remained in Bethany, Tecumseh, and Aliquippa, but all production eventually consolidated at Bethany and the new Albany, Georgia, facility.[16] By the end of 1966 Rockwell-Standard had acquired Aero Commander, Meyers, CallAir, and Volaircraft, plus the small Snow Aeronautical of Olney, Texas. Leland Snow, a veteran agricultural pilot, began development of original aircraft in 1955 and incorporated in 1961. His first design, the S-2B, gained certification in 1958. With sales of more than 200 agricultural aircraft by 1964, Snow had attained a strong position in that segment by the time of the merger. Earlier, Rockwell-Standard had made unsuccessful merger overtures to Beech, Cessna, Mooney, and Piper.[17]

Fletcher Aviation Corporation, in addition to its production of aircraft components and jettisonable fuel tanks, and subcontracting on military missiles, continued to develop small aircraft. After the FD-25 Defender, the firm developed the FU-24 Utility in July 1954, which was produced under license in New Zealand from 1957 as an agricultural aircraft. The firm reorganized in 1960 as the Flair (from Fletcher Aircraft) Aviation Company, though still managed by Wendell Fletcher. Then Flair was acquired by American Jet Industries, founded in 1951 for aircraft modification and repair, and renamed the Sargent-Fletcher Company in 1964, with the participation of E. J. Sargent. The company continued its other lines, but the FU-24 was its final aircraft development, and all rights were sold to New Zealand in 1964.

One smaller venture which was to endure was that of Ohio native Belford D. Maule. An early experimenter who had built his first airplane in 1931 and later worked for Lycoming, Maule began an aircraft parts business in Jackson, Michigan, in 1940. As the B. D. Maule Company, he designed a small four-seat high-wing model named the Bee Dee M-4 and based somewhat on the old Piper Cub. It first flew in 1957. The further developed M-4 prototype first flew on September 8, 1960, and formation of the Maule Aircraft Corporation followed in 1961. The M-4 Rocket, a rugged utility model with STOL characteristics, went into production in 1962. An improved model with all-metal wings followed in 1963, and sales gradually rose.

Ted Smith departed Aero Commander in 1963 and founded Ted Smith Aircraft Company, Inc., in California to develop a new design. His Aerostar Model 600 twin-engined executive transport with a mid-wing configuration, unusual in a general aviation aircraft, first flew in November 1966 and led to a series of developments. The Aerostar, while matching the performance of other modern twin-engined business aircraft, was designed with simpler features and fewer parts for ease of construction and maintenance. Smith added the Model 601 with turbocharged engines, but would encounter persistent problems in establishing the Aerostar in production.

The engineer Edward J. Swearingen, who earlier had worked for Lear, established Swearingen Aircraft in San Antonio, Texas, to develop and market modifications of existing light twin aircraft for business and commuter purposes. Swearingen then developed the piston-engined SA-26 Merlin I, which used components from the Beech Twin Bonanza and Queen Air but was essentially a new design. Then the turbine-powered SA-26T Merlin II, first flying on April 13, 1965, marked the complete transition of Swearingen from modification to manufacturing. With rising business, Swearingen undertook major factory expansion for production of both models, which led to a long series of developments.

Lockheed again explored the general aviation sector with a rugged single-engine light utility aircraft for use primarily in less-developed areas of the world. Designed by the Mooney brothers, the LASA-60 was built at the Lockheed Marietta, Georgia, factory and first flew on September 15, 1959. Lockheed did not plan to produce the LASA-60 domestically, instead licensing production to the Lockheed-Azcarate affiliate formed in Mexico and to other Lockheed affiliates in Italy and Argentina. The program was less than a resounding success, however, as only 18 Mexican models were built, all for the Mexican

Air Force. And the Lockheed-Kaiser venture in Argentina was abandoned. Macchi of Italy retained exclusive rights but produced only 100, ending the program in 1965.

Fairchild, with declining military production, entered the utility aircraft market with licensed production of the Swiss Pilatus Porter, a STOL design powered by a small turboprop engine. Initially named the Heli-Porter, it was renamed Turbo-Porter and entered production in 1966 at the Hiller Aircraft Division, after Fairchild acquired Hiller.

Among experimental firms, M. B. Taylor developed the Aerocar Model II Aero-Plane, a nonroadable version of the Aerocar. Taylor also refined the basic Aerocar, which won CAA certification in 1956. A market, however, still did not develop; only six were sold. Willard R. Custer developed aircraft using his channel-wing concept, in which pusher engines would pull air over the channeled wing sections. Custer claimed much greater lift would be developed than with a conventional airfoil, giving the aircraft capabilities approaching those of the helicopter, including near-hovering speeds. After years of experiments, Custer first tested his CCW-5 business aircraft prototype, built by Baumann in California and modified from the Brigadier, on July 13, 1953. The Custer Channel Wing Corporation then was formed in Hagerstown, Maryland, in 1956 to produce the CCW-5 and to pursue further developments. Custer rented hangar space earlier owned by Kreider-Reisner and Fairchild.

Production encountered numerous delays and financing problems. Plans were announced variously for production at sites in Texas and Canada, but never materialized. Further, widespread skepticism remained that the channel wing would attain the capabilities claimed for it. What was announced as the first production CCW-5 finally flew on June 19, 1964, at Hagerstown, but a long dispute with the Securities and Exchange Commission over public stock sales and failure to gain FAA certification effectively killed prospects.

Light Helicopter Development

Development of general aviation helicopters progressed during the 1954–67 period. While distinctly subordinate to the military market, the civil helicopter market gradually expanded after the Korean War, with increasing applications in agricultural spraying, law enforcement, geological and forestry survey, medical transportation, and firefighting. But the helicopter as an executive transport had yet to become established.

A new civil design was that of R. J. Enstrom, who founded Enstrom Helicopters in 1959. Enstrom's Model F-28 ultralight two-seat design first flew on December 12, 1960. A three-seat development first flew on May 26, 1962, but did not enter production until 1966. The Brantly B-2, also in development for years, finally entered production in 1959 and more than 200 were produced. It was succeeded by the developed Model B-2A, and a larger five-seat Model B-305 appeared in 1965. From the late 1950s Cessna produced small numbers of its YH-41 Seneca development of the original Siebel helicopter for the Army, as well as a few of the CH-1 civil version, but poor sales led to discontinuation of the line in 1962.

The most successful general aviation helicopter continued to be the Bell 47, which underwent continuous improvement. The Model 47H, an updated enclosed-cabin version, appeared in 1954 but did not win a wide market. Then the Model 47J-2A Ranger, a deluxe four-seat development with a streamlined monocoque fuselage, was introduced in 1956 and gained popularity both in civil and military roles. Bell Aerospace, including its Bell Helicopter operating unit, became a wholly owned subsidiary of Textron, Inc., on July 5, 1960.

An event with significant implications for small civil helicopters was the important Army competition in the early 1960s for a light observation helicopter (LOH) to replace the Bell H-13 variants, the Hiller H-23, and the Cessna L-19 (O-1). The potential market was estimated to involve several thousand helicopters. Howard Hughes, among his numerous aviation interests, had entered the light helicopter field by forming Hughes Helicopters as a division of his Hughes Tool Company. The first Hughes-designed helicopter was the two-seat Model 269 of 1955, for both military training and civil roles. Hughes, Bell, and Hiller were the finalists for the LOH order, won by the developed Hughes Model 500 as the OH-6A. Hughes hoped to capture a worldwide civil market as well. The Model 269, redesignated Model 200, and the larger Model 500 were produced for that market. Three hundred Model 200s were sold by 1964.

Stanley Hiller negotiated a merger of his company, again renamed as Hiller Aircraft Corporation, into Electric Autolite Corporation of

Ohio, a longtime business aircraft user, in the summer of 1960. But the autonomous Hiller Helicopters subsidiary continued to lose money. Then Fairchild, in a diversification move sparked by the prospect of the lucrative Army LOH contract, acquired Hiller Helicopters from the ELTRA Corporation, renamed from Electric Autolite, in May 1964. The company became Fairchild Hiller Corporation in September, with Stanley Hiller becoming an executive vice president of the corporation in addition to heading helicopter operations.[18] Helicopter production remained in California.

While its OH-5A entry in the LOH competition was unsuccessful, the Hiller Aircraft Division carried forward with the turbine-powered FH-1100 civil development. The FH-1100 first flew on January 26, 1963, and entered production in 1966. Stanley Hiller eventually left Fairchild Hiller to return to California and other business interests.

Bell's Model 206A JetRanger light turbine-powered design, also unsuccessful in the LOH competition, first flew as a civil helicopter on January 10, 1966, and entered service in 1967. It led the market in its category and also spurred a general expansion in business use of helicopters.

A sad footnote to an era of rapid rotary-wing progress was the death of autogiro pioneer Harold Pitcairn on April 23, 1960, at age 62, by his own hand.

# # # # #

General aviation continued as a growth market through most of the 1950s and 1960s. By 1964, 40 percent of the general aviation fleet was more than 15 years old, extending largely from the short-lived production boom of 1946–1947.[19] The requirement for replacement of aging aircraft supported continuing growth. As with military aircraft, piston-engined general aviation aircraft reached a broad performance plateau during this period but continued to grow in cost and complexity. General aviation had become a key component of the broad aerospace industry, but it still remained by far the smallest sector. By 1967, for example, McDonnell Douglas and Boeing were approaching $3 billion in annual sales each, while the largest general aviation firms, Cessna and Beech, posted sales of only $213 million and $174 million, respectively.[20] Furthermore, both totals reflected major military components.

But expectations for continued growth and prosperity were the order of the day. Cessna, for example, had forecast annual production of 20,000 general aviation aircraft by 1970, the majority being piston-powered single-engined models.[21] Larger business and executive aircraft continued as a major growth segment, as did specialized agricultural aircraft. Development of chemical sprays enhanced agricultural aircraft market growth to the point that there were some 5,000 agricultural aircraft in operation by 1966. Air taxi and commuter services using multiplace aircraft also grew rapidly during the 1960s.

The United States enjoyed a strong export market in general aviation aircraft, as no effective international competition existed in the field. In 1966, for example, 15,700 general aviation aircraft were produced in the United States, of which 20 percent were exported. The British, with their wartime debt burden, had embargoed light aircraft imports from the end of the Second World War until 1960. When the embargo was lifted, American aircraft quickly became a factor.

The Business Jet

Corporate executive transportation became generally established during the 1950s, and strong growth continued. The rapid growth and increasing diversification of large corporations, with geographically dispersed facilities, led to corollary growth in corporate air fleets. Business aircraft use also was spurred by the view that scheduled airlines did not offer the timeliness, flexibility, and comprehensive route structures increasingly required by business. Manufacturers found that business aircraft could be effectively marketed on their capabilities to visit several sites in one day, impossible on scheduled airlines and enormously beneficial to busy executives and sales teams.

While large corporate aircraft were widely used, before the 1960s most were conversions of smaller airliners such as the DC-3 and Convair 340, or of Second World War-vintage bombers such as the Douglas B-23, Lockheed Model 18, and Martin B-26. Then with the advent of large jet-powered airliners, small business jets became increasingly regarded as feasible. But in that emerging market the largest general aviation firms, Beech and Cessna, faced a potential threat from large military manufacturers whose capabilities might be used to great advantage in business jets. Several military firms in fact designed specialized corporate jet transports to replace the mix of propeller-driven smaller business and converted military and commercial aircraft. Business jet development was to influence significantly the future structure of the general aviation industry.

The Grumman Gulfstream was an early and successful specialized corporate transport. A further manifestation of Grumman's diversification strategy, the G-159 Gulfstream first flew on August 14, 1958, and led to a long series of Grumman corporate designs. Almost of small airliner size, the Gulfstream was powered by two Rolls-Royce Dart turboprops. It represented a considerable risk for Grumman, with an initial price in the $1 million range. But the Gulfstream gained market popularity as well as a reputation as the Rolls-Royce of business aircraft. After production on Long Island, Grumman opened a new Savannah, Georgia, factory specifically for the Gulfstream on September 29, 1967.

The French were the first to produce a turbojet business or executive aircraft, the small four-seat Morane-Saulnier MS.760 Paris of 1954, itself developed from a jet trainer. Beech acquired American rights to the MS.760 in 1955 but decided against marketing or production.[22] Cessna tested a comparable business jet developed from its T-37 trainer but likewise did not proceed with production. The United States later captured the world lead in business jets, although other nations would gain and maintain a significant market share.

Major spurs to development of business jets were the Air Force UTX and UCX competitions beginning in 1956. The UTX and UCX involved small jet passenger aircraft for military training and utility transport missions, respectively, but which possessed obvious potential for business use. Development of economical and reliable small turbofan engines such as the Pratt & Whitney JT12 and the General Electric CJ610 also facilitated business jets. The Lockheed JetStar, the North American Sabreliner, and the McDonnell Model 119 were developed originally for the UTX/UCX competitions.

Fairchild did not continue its M185 project after it lost the Air Force competition, however, and McDonnell's Model 220, the business market counterpart of the Model 119, also did not proceed.

The Lockheed JetStar, to which the Mooney brothers had contributed, first flew on September 4, 1957, and won the UCX competition as the C-140. The smaller Sabreliner, which first flew on September 16, 1958, won the UTX competition as the T-39. The base of military orders supported development of both as business jets, the first to attain production. Entering operation in 1962, the JetStar became the top-line business jet until the arrival of the Gulfstream II, but orders for civil JetStar and Sabreliner models were comparatively slow in building. Both were still regarded as too expensive for all but the largest corporations. Real growth came with the LearJet.

The inventor and aviation entrepreneur William P. Lear began planning a small executive jet in 1959, to be named the LearJet. Lear, a high school dropout who had made a fortune in aviation instruments and electronics, held some 150 aviation-related patents. Lear, Incorporated, earlier had marketed conversions of the Lockheed Lodestar as the Learstar luxury executive transport, but the venture was not profitable. Lear first approached Mitsubishi of Japan to produce a LearJet but did not succeed. The Lear board of directors also declined to finance the venture, whereupon Lear sold his interest in 1959 in order to pursue the project on his own.[23] Lear proceeded without surveys of market demand for the jet, relying solely on his intuition. (Lear, Incorporated, later merged with Siegler to become Lear Siegler Corporation.)

William Lear moved to Switzerland, where he held existing business interests, and established the Swiss American Aviation Corporation (SAAC) to develop the LearJet. Originally designated SAAC-23, the design used the wing and basic structure of the experimental Swiss AFA P-16 strike fighter then being tested. Lear felt at the time that he could better control costs in Switzerland, but numerous problems with suppliers led him to move the project, including production tooling, back to the United States in 1962. Lear also benefited in design of the LearJet by NACA/NASA-developed design technology, as he lacked resources for basic research.[24] He located his renamed Lear Jet Industries in Wichita, where he won economic incentives as well as access to a large pool of experienced labor. Cessna's T-37 program had created a reservoir of production skills in small jet aircraft, and Wichita issued the first industrial revenue bonds in its history to attract Lear. Lear Jet Industries was, incidentally, the first firm to be formed specifically for the business jet market.

The LearJet Model 23, configured for six passengers and powered by twin rear-mounted General Electric CJ610 turbojets, first flew on October 7, 1963. Lear had staked his entire $10 million fortune on the venture but soon ran dangerously low on capital as he pressed to beat the rival Jet Commander to the market. The first delivery to a customer took place on October 13, 1964, and soon thereafter, on November 30, 1964, a successful public stock offering provided required capital. At an initial price of $540,000 the LearJet gained instant popularity, with over 100 sold by the end of 1965. The improved and refined Model 24 replaced the Model 23 in production in March 1966 and was joined by the larger Model 25, first flown on August 12, 1966.[25] An around-the-world LearJet flight, completed on May 29, 1966, garnered much favorable publicity for the design. The flight took only 65 hours.

In an expansionist period, Lear even contemplated briefly development of a 40-seat feederliner based on the LearJet configuration. Then in May 1966, Lear Jet acquired Brantly Helicopter Corporation, still trying to perfect and market a two-place, simple, affordable helicopter. Newby O. Brantly accompanied the move, but continued financial losses forced a later sale of the helicopter operation.

Lear, in common with many industry pioneers, was a better designer and engineer than corporate executive, and after initially strong LearJet sales, the firm experienced a sales decline as well as major difficulties with a haphazard distribution network. Overdiversification, operating losses, and increasing competition led to a financial crisis, requiring a rescue by merger. Cessna, planning to develop its own business jet, was uninterested. Beech, favored by several Lear executives as a logical candidate, declined primarily due to longstanding differences between Lear and Frank Hedrick.[26] Then in April 1967, Charles C. Gates, head of Gates Rubber Company in Denver and a veteran pilot, agreed to purchase Lear's 62 percent holding for $21 million. His Gates Aviation Corporation, a sales and service company, initially became a subsidiary of Lear Jet Industries, but the two eventually merged as the Gates LearJet Corporation. The future of the design seemed assured under new ownership.

Rockwell Aero Commander entered the executive jet market with the Jet Commander, chiefly designed by Ted Smith and directly competitive with the LearJet. Also powered by twin rear-mounted GE CJ610s, but with a mid-mounted wing, the Jet Commander actually preceded the LearJet in the air, making its first flight on January 27, 1963. Entering service in 1965, it also attracted initially strong orders.

A Douglas business jet, the PD-808, was designed by the El Segundo Division in cooperation with the Italian firm Piaggio. The first prototype flew on August 29, 1964. Piaggio was licensed to produce the PD-808 for both civil and military uses, but it eventually entered service only with the Italian Air Force. Douglas retained but did not exercise the U.S. manufacturing and marketing rights.

With the LearJet, Jet Commander, Sabreliner, and JetStar already on the market, Grumman followed its turboprop Gulfstream I with the Gulfstream II, an entirely new swept-wing design powered by Rolls-Royce Tay turbojets. It first flew on October 2, 1966. It was also produced at Savannah and became an immediate success. The largest in the field, the Gulfstream II was regarded as the ultimate in executive jets and led to further developments.

# # # # #

Roughly concurrent with the development of executive jets was the development of so-called third-level, or feeder, airliners, which also carried significant implications for general aviation manufacturers. The term *third-level airline* began to be used in the early 1960s to distinguish that category from major trunk carriers and regional or local service carriers, and third-level service steadily expanded throughout the country and the world. This led to a requirement for modern small airliners reminiscent of the Lockheed Saturn and Beech "Twin Quad" designs of 1946 and 1947.

Rule changes by the Federal Aviation Agency (FAA), which had succeeded the CAA on December 31, 1958, facilitated development. Federal Air Regulation (FAR) Part 23, dealing with aircraft under 12,500 pounds gross weight, was less stringent than regulations for larger aircraft, making commuter operation more feasible. Moreover, single-pilot operation, virtually an economic necessity at that level, was restricted by FAR Part 298 to aircraft carrying no more than 19 passengers and weighing no more than 12,500 pounds. But those carrying 20 or more passengers required radar and complete instrumentation, and consequently more regulation, covered by FAR Part 25.

Air taxi operators were exempted from full airline regulation and permitted to operate scheduled services as long as they stayed within the weight limits. The exemption was extended by the FAA indefinitely in 1965. One sidelight to these developments was that the distinction between larger general aviation aircraft and commercial aircraft became less distinct. General aviation aircraft had been classified as those with capacities of 19 seats or fewer and under 33,000 pounds gross weight, but such aircraft as the Gulfstream tended to push the definition.

Commuter airline growth was also aided by the trend of trunk airlines away from service to smaller communities, leaving their limited passenger markets to the smaller lines. Adaptations of business jets were regarded as unfeasible, given their cost and the economics of short-range operation, but general aviation firms readied specialized designs for the commuter market, generally with a twin-turboprop configuration.

6

The Modern Era, 1967–1979

General aviation had made significant technological advances during the 1950s, but as the industry matured during the 1960s, its products became rather similar in design and performance. Light aircraft technology continued largely static during the 1970s, and the old dream of an affordable mass-market aircraft had long been abandoned.

A particular problem in general aviation design was that requirements were not spelled out precisely by potential customers as was the case with military and commercial aircraft. Accordingly, manufacturers did not develop new designs from customer specifications but simply tried to guess which would be most likely to sell. In marked contrast to relatively static piston-powered aircraft technology was the intensive development of larger turbine-powered executive transports, particularly business jets. Business flying dominated general aviation, with 72 percent of the general aviation fleet devoted to business use (see Table 6-1).[1]

Personal and recreational flying remained limited by price, but business users were more willing to pay for technological advances. One advance from the 1970s was that single-engined business aircraft were increasingly offered with turbocharged piston engines, enabling them to compete in speed and altitude performance with turbine-powered models. Cessna, Piper, Beech, and Mooney eventually offered single-engined turbocharged models. Another significant benefit was that the National Aeronautics and Space Administration (NASA, successor to NACA) began to devote more research to general aviation, specifically in such design innovations as canard wings, simulation and analytical techniques,

Table 6-1
Annual Production and Sales, General Aviation Aircraft

Year	Units	Factory sales ($ millions)
1968	13,698	425.7
1969	12,457	584.5
1970	7,262	337.0
1971	7,466	321.5
1972	9,774	557.6
1973	13,646	828.1
1974	14,166	909.4
1975	14,056	1,032.9
1976	15,451	1,225.5
1977	16,904	1,488.1
1978	17,811	1,781.2
1979	17,048	2,165.0
1980	11,877	2,486.2

SOURCE: *General Aviation Statistical Databook* (1990–1991 edition), p. 4.

laminar flow airfoils, and drag reduction; in such safety-related areas as stall/spin phenomena; and in composite materials and weight reduction.[2]

Even as overall flying safety improved, a conflict over safety between business and commercial flying emerged and posed another threat to general aviation growth. Several widely publicized collisions between smaller business aircraft and airliners in the late 1960s led to allegations that private pilots were less skilled and less cautious than commercial pilots, and that increasing business aircraft traffic threatened commercial air safety. Investigation generally concluded that the crashes owed as much to commercial pilot error as to private pilot error, and the controversy gradually eased. One conclusion, however, was that the overburdened and antiquated air traffic control (ATC) system needed major upgrading. Increasing use of business airports in major cities eased the strain on commercial airports, also alleviating safety concerns.

The Aerospace Industries Association (AIA, renamed in 1959), which had represented general aviation interests through its Utility

Airplane Council since the 1950s, saw those interests spun off into an independent organization. The General Aviation Manufacturers Association (GAMA) was formed in January 1970 to advance the cause of general aviation and to enhance the environment for its growth. The mature general aviation sector then was represented by NBAA, AOPA, EAA, and GAMA. General aviation aircraft sales exceeded $1 billion for the first time in 1975.[3] Time took its toll on light aircraft pioneers during this period, however, as Rae Rearwin died on November 16, 1969, at age 91; Lloyd Stearman died on April 4, 1975, aged 76; and Clayton Brukner of Waco died on December 26, 1977, at 81. Alfred Verville died in 1970, and Sherman Fairchild in 1971.

Industry Survey

Production boomed through most of the 1960s (see Tables 6-2 and 6-3), but suffered a sharp drop during the recession of 1970–1971. Wichita was especially affected, as area unemployment rose sharply. But shipments and dollar value of sales began to rise again in 1972 and increased through the remainder of the decade as general aviation growth continued. Prosperity was not spread evenly through the sector, however, as most smaller firms still struggled to compete against the Big Three of Beech, Cessna, and Piper. But Mooney maintained a niche in the four-place, low-wing, retractable-gear segment, as did Rockwell Aero Commander in twin-engine executive transports.

The period also was characterized by frequent company reorganizations and ownership changes, and long-term stability remained elusive. There was a trend toward acquisitions of general aviation manufacturers by larger corporations seeking diversification, with both aerospace and nonaerospace firms taking positions in the field.

Product development was especially active in major firms. The new North American Rockwell Corporation (NAR), formed in November 1967, combined its then-extensive general aviation aircraft operations, all from acquisitions, into the Commercial Products Group, which also encompassed most activities of the predecessor Rockwell-Standard Corporation in Pittsburgh. Most general aviation production was controlled by the Aero Commander Division, with plants in Bethany, Oklahoma; Albany, Georgia; and Olney, Texas.

Table 6-2
Shipments of General Aviation Aircraft by Selected Manufacturers, 1947–1968

Year	Total	Beech	Cessna	Champion	Lear	Lockheed	Mooney	Piper	NAR*	Other
1947	15,764	1,288	2,390	—	—	—	—	3,634	—	8,452
1948	7,037	746	1,631	—	—	—	—	1,479	—	3,181
1950	3,386	489	1,134	—	—	—	51	1,108	—	604
1952	3,058	414	1,373	—	—	—	49	1,161	39	22
1954	3,071	579	1,200	—	—	—	14	1,191	67	20
1956	6,843	724	3,255	162	—	—	154	2,329	164	55
1958	6,414	694	2,926	296	—	—	160	2,162	97	79
1960	7,588	962	3,720	248	—	—	172	2,313	155	18
1961	6,797	818	2,746	112	—	—	286	2,646	139	50
1962	6,723	830	3,124	91	—	9	387	2,139	121	22
1963	7,603	1,061	3,456	99	—	10	502	2,321	114	40
1964	9,371	1,103	4,188	60	3	6	650	3,196	109	56
1965	11,967	1,192	5,629	271	80	18	775	3,776	110	116
1966	15,747	1,535	7,888	331	51	24	917	4,437	354	210
1967	13,577	1,260	6,233	267	34	19	642	4,490	386	246
1968	13,698	1,347	6,578	255	41	16	579	4,228	471	183

*Includes Sabreliner and original Aero Commander Company.

SOURCE: *Aviation Facts and Figures, 1969,* Aerospace Industries Association, Washington, D.C., 1969, p. 34.

Table 6-3
Production of General Aviation Aircraft by 14 Manufacturers, 1968*

Manufacturer	Number	Billing price (000)
Aero Commander (NAR)	435	$ 22,309
American Aviation	33	291
Beech	1,347	115,737
Bellanca	94	2,128
Cessna	6,578	138,784
Champion	255	2,248
Lake	30	801
Lear Jet	41	28,650
Lockheed	16	N/A
Maule	25	332
Mooney	579	24,707
North American Sabreliner	36	N/A
Piper	4,228	85,484
Ted Smith	1	71
General aviation total	13,698	$421,522**

*Excludes military sales, helicopters, and gliders.

**Excludes North American Sabreliner and Lockheed JetStar.

SOURCE: *Aviation Facts and Figures, 1969,* Aerospace Industries Association, Washington, D.C., 1969, pp. 31–33.

All aircraft in the Aero Commander line were given bird names. The Volaire 1050, first redesignated the Aero Commander 100, reflecting its new affiliation, subsequently was renamed the Darter Commander. From the Darter Commander was developed the Lark Commander, a refined, more luxurious version. But sales of the former Volaire 1050 line were below expectations. The Aero Commander 200, originally the Meyers Model 200D, a low-wing, four-place model dating from 1953, also experienced poor sales, and production was ended and design rights sold.

North American Rockwell continued production of the established Aero Commander twin-engined line, including the Shrike Commander, Courser Commander, and Turbo II Commander, later renamed Hawk Commander. Antitrust considerations arising from the North American merger required the divestiture of either the Sabreliner or Jet Commander business jets. Subsequently, all rights to the Jet Commander were sold to Israeli Aircraft Industries in 1968 after 149 had been built. In the extensive agricultural aircraft line, the original CallAir A-9 was renamed Sparrow Commander and was joined by the Quail Commander, with an uprated engine, and then by a further upgraded Snipe Commander. The new Thrush Commander, from the Snow acquisition and the largest agricultural aircraft on the market, completed the line.

North American Rockwell further reorganized on September 30, 1969, with the Commercial Products Group succeeded by an Industrial Products Group. The Industrial Products Group controlled all general aviation production, marketing, and other services within a new General Aviation Division. Departing from its practice of adapting existing designs, the division developed the new four-seat low-wing Aero Commander 112, announced in December 1969, which went into production in 1972. The General Aviation Division by that time offered the Lark Commander, Aero Commander 112, Shrike Commander and Shrike Commander Esquire, the Turbo Commander 681B (again renamed), plus the agricultural line. The Sabreliner was also included in the new division, and NAR thus offered almost comprehensive coverage of the general aviation market. A new piston-engined Aero Commander 685 was developed from the Turbo II Commander. The Model 112, renamed Alpine Commander, was joined by the Model 111 fixed-gear variant and by the deluxe Gran Turismo Commander.

Despite strong development activity, NAR's general aviation production remained unstable. The company sold rights to the Thrush Commander and Quail Commander models in 1971 to a new Mexican operation in which it held 30 percent ownership. Mexican production was troubled, however, and the venture soon ended. Lark Commander production ended in 1972, and remaining general aviation manufacturing other than the Sabreliner was consolidated at the Bethany plant in 1974.

In 1971 Leland Snow formed a new Air Tractor, Inc., to continue independent development and production of large agricultural aircraft. The company remained in Olney, Texas.

Renamed Rockwell International in 1973 and embarking on a strategy of reducing its aviation interests, the company sold its final Thrush Commander 600 and 800 series to the new Ayres Corporation in November 1977. Ayres operated from the former Rockwell facilities at Albany, Georgia. Of the original Aero Commander twin-engined line, only the refined Turbo Commander 690B, which succeeded the Turbo Commander 681B in 1972, remained in production through 1979. Production of the single-engined Alpine Commander and Gran Turismo Commander models ended in September 1979, but the advanced Aero Commander 840, which first flew on May 17, 1979, succeeded the earlier Turbo Commander. By that time, however, Rockwell faced dim prospects in the overcrowded market and planned to exit general aviation entirely.

Although completely unrelated, Mooney and Aerostar would become associated by acquisition. Mooney Aircraft, Inc., acquired Alon, Inc., which held rights to the original Ercoupe design, on October 10, 1967. The Alon A-2A Aircoupe was added to the Mooney line, joined by the refined, single-tail M-10 Cadet model. The program was transferred to the Texas factory, but few were produced, ending the long production history of the Ercoupe. Mooney models continued, and the popular Mark 21 was renamed the Ranger. Mooney also progressed from marketing to assembly of the Japanese-designed MU-2 twin-turboprop business aircraft at its San Angelo factory. Mooney then was acquired by American Electronic Laboratories of Pennsylvania on March 26, 1969, and operated as Mooney Aircraft Corporation. This arrangement was short-lived, however, as Butler Aviation International of Texas bought Mooney from A.E.L. on November 21, 1969, operating it as a 100 percent-owned subsidiary.

The Aerostar design of Ted Smith experienced a checkered ownership and production history. Rockwell-Standard, just prior to the North American merger, allowed its purchase option on Ted Smith Aircraft Corporation to expire in April 1967. Ted Smith then moved independently to open a new factory for production of the Aerostar on January 5, 1968, in Van Nuys, California, but persistent financing difficulties slowed deliveries. Smith later sold control to American Cement Corporation. Then Butler Aviation International, which already owned Mooney, acquired Aerostar on February 16, 1970. The subsidiary, combining both lines, was renamed Aerostar International

Corporation, and production attained some stability. Aircraft were marketed under the Aerostar nameplate, with four original Aerostar series employing the same basic airframe. The subsidiary name was again changed, this time to Aerostar Aircraft Corporation, on July 1, 1970, and Mooney designs were marketed under the Aerostar nameplate as well. Market weakness then led Aerostar to suspend all production in 1972.

Ted Smith, having lost his company earlier, saw an opportunity to regain control and formed a new venture, Ted Smith and Associates. Reacquiring Aerostar components from Butler International at a distress price, he restarted production in Santa Maria, California, and planned more advanced developments. The venture became Ted Smith Aerostar Corporation, but Ted Smith died in 1976 and the company later was acquired by Piper. In the meantime, Republic Steel Corporation of Cleveland (not to be confused with the old Republic Aviation) announced its acquisition of the Mooney line from Butler on October 14, 1973. The deal became effective early in 1974, recreating Mooney Aircraft Corporation in the process. The Mooney line again enjoyed success in the market. The lengthened M-20F Executive led to the improved Model 201 and turbocharged Model 231 of 1976, largely designed by LeRoy LoPresti. The lower-priced Ranger also continued.

Piper enjoyed overall prosperity during the 1960s, but remained the smallest of the Big Three, possessed the smallest military segment, and offered the narrowest product line. The company delivered its 10,000th Cherokee in August 1967. Traditionally, Piper had competed with simplicity and affordability, but it was increasingly pressured by Beech, Cessna, and Rockwell. The company suffered a sharp profit drop during the 1967 market downturn, largely due to a lack of broad market coverage. It faced a growing problem in not having a contender in the small commuter airliner market to compete with the Beech 99 and others. Product development remained active, but the six-eight seat PA-31 Navajo did not enter service until April 1967, later than the competitive Cessna 411 and Beech Queen Air. An improved PA-39 Twin Comanche appeared in 1971, but the Twin Comanche line ended in 1974. Single-engined production, down to the veteran Super Cub, remained strong.

The initial commuter airliner entry, and the largest aircraft ever designed by Piper, the 15-seat PA-35 Pocono, first flew on May 13,

1968. Production was planned for a new factory in Lakeland, Florida. William T. Piper, Sr., dubious of turboprop power for smaller aircraft, powered the Pocono by turbocharged piston engines instead. He felt that the price advantage of a piston-powered, unpressurized aircraft would enable it to capture a market from far more expensive turbine-powered, pressurized competitors.[4] The Pocono proved to be seriously underpowered, however, and production was shelved in 1970, still leaving Piper without a design for the commuter market.

Bill Piper, Jr. finally succeeded his father in the office of president on February 6, 1968, although the senior Piper remained chairman of the board. Younger brothers Tony and Pug Piper continued as executives. By 1969, with the Piper family holding only 31 percent of the company stock, sales approaching $100 million annually, and a strong cash position and strong profits, the firm was an attractive acquisition candidate.[5] On January 23, 1969, Herbert J. Siegel, head of Chris-Craft, Inc., known as a boat manufacturer but actually a highly diversified company, announced the acquisition of a substantial block of Piper stock with the intent to acquire control. It tendered for 300,000 shares in the open market. Bill Piper resisted the takeover attempt, and a protracted battle ensued.

The Pipers first approached Grumman about acquiring control, but discussions eventually failed. Piper finally found a corporate rescuer in the Bangor Punta Corporation of Greenwich, Connecticut, which acquired the more than 500,000 shares held by the Piper family plus others to constitute 51 percent to Chris-Craft's 42 percent.[6] The transaction, at the end of 1969, netted the Piper interests more than $30 million, but control passed from the family's hands. Chris-Craft pursued a legal battle against Bangor Punta, but unsuccessfully. After the takeover the senior Piper resigned as chairman of the board and director. Tony and Howard Piper and Walter Jamouneau also resigned from the board, but remained officers of the company. William T. Piper, Jr. remained president and a board member.

William T. Piper, Sr. died on January 15, 1970, at the age of 89, after a 40-year career in the industry. In September 1970, Bill Piper, Jr. relinquished the presidency to Joseph T. Mergen, previously vice president and general manager of Avco-Lycoming and earlier executive vice president of the Curtiss-Wright Corporation.[7] Tony Piper soon moved to Texas and severed all ties with the company. In addition to the managerial upheaval, the general market decline at the time affected Piper more severely than others and forced the temporary closure of both the Lock Haven and Vero Beach plants beginning in the summer of 1970.

In a second attempt to enter the commuter market, Piper discussed acquisition of privately held Swearingen Aircraft, makers of the successful Metro, in 1971, but the deal did not reach fruition.[8] Success came in the higher end of the executive market with the first Piper-designed turboprop, the PA-31T Cheyenne, which first flew on August 20, 1969. Derived from the Navajo, and first converted by Swearingen, the Cheyenne (later Cheyenne II), after some delay, finally went into production in 1973 at the new Lakeland plant. Piper also introduced the smaller PA-34 Seneca (originally Twin Cherokee), eventually to succeed the older Twin Comanche, on September 23, 1971. The PA-31-350 Chieftain, a lengthened Navajo with up to 10 seats was offered from 1972 for the commuter market. Further single-engined developments introduced concurrently in 1970 were the Cherokee Arrow, with retractable gear, and the Pawnee II agricultural model. The larger and more advanced PA-36 Brave joined the Pawnee II in 1974.

Bill Piper, Jr. left the board and Howard "Pug" Piper resigned as an officer in January 1973, formally ending involvement of the Piper family. Mergen succeeded in turning the company around by the end of 1973, with strong sales and profits returning, but then resigned in early 1974 after a series of health problems and frustration with conflicts between the Bangor Punta and Chris-Craft interests.[9] He was eventually succeeded by J. Lynn Helms, formerly of United Aircraft and later chairman of the FAA. Helms pushed ahead with twin-turboprop executive transports, commuter models, and modern single-engined designs. Piper once again became a strong competitor in other than the business jet segment. The firm reached the production milestone of 100,000 aircraft in April 1976.

Diversification efforts continued. Piper had always lagged Beech and Cessna in military business, and in 1971 it purchased the rights to the Cavalier Mustang II, a modified, turboprop-powered version of the famed P-51 Mustang fighter. Trans-Florida Aviation, owned by entrepreneur David B. Lindsay, Jr., developed the Mustang II and proposed it to the military as a low-cost close-support aircraft. Piper moved the program, renamed the Piper PA-48 Enforcer, to its Vero Beach plant. Piper's marketing efforts failed, however, as the Air Force maintained it had no requirement. But Piper suspected that the Air Force was more interested in protecting its investment in the more complex and expensive Vought A-7 Corsair II and Fairchild A-10 Thunderbolt II.[10]

While unsuccessful in the military market, Piper expanded its international activities, concluding licensing agreements with Argentina in 1971, Brazil in 1974, and Poland in 1977. Brazil in particular had been a major general aviation export market during the 1971–1974 period, but the market had disappeared rapidly as Brazil, protecting its domestic aircraft industry, imposed a high tariff barrier for aircraft of under 15,400 pounds. Brazil was the most active of the international operations, with Embraer, Piper's licensee, undertaking complete manufacture rather than assembling aircraft from imported components, although some subassemblies were imported. Piper thus was able to continue serving the market, while Cessna was effectively frozen out. As was frequently encountered in new foreign licensed production, aircraft built in Brazil cost up to 50 percent more than those made in Vero Beach, but the cost premium later dropped to 25 percent with efficiency gains.[11] PZL of Poland produced the PA-34 Seneca II under license, while the Argentine and Brazilian operations soon cooperated in production and marketing.

On March 27, 1978, Piper acquired Ted Smith Aerostar Corporation, which it operated initially as the Santa Maria Division. Further developments included the longer, more powerful PA-42 Cheyenne III of 1977 and the PA-38 Tomahawk, a small two-seat trainer intended to challenge the Cessna 150/152, in 1978. The Cheyenne III proved to be a viable competitor to the turbine-powered Cessna Conquest and Beech King Air, and the simpified Cheyenne I, with an attractive base price of $600,000, also sold well.

The basic four-seat, low-wing PA-28 line originating with the Cherokee grew to include the Warrior, Archer, Arrow, and Dakota. The larger six-seat PA-32 line included the Cherokee Six, Lance 2, and more powerful Turbo Lance 2, some with fixed and others with retractable gear. The T-tail configuration was adopted for many Piper aircraft during this period. The new PA-32 Saratoga, a roomier, more modern executive model developed from the Cherokee Six, appeared in 1979 and succeeded earlier six-seat models.

With continuing production of such established models as the Super Cub, Aztec F (introduced in 1978), Navajo, Cherokee, Cheyenne, Chieftain, and Pawnee, plus the Aerostar line, Piper finally approached the comprehensive market coverage of its major competitors. Product proliferation approached that of Cessna. Bangor Punta purchased the Chris-Craft shares in 1977, ending the control question, and Piper enjoyed a strong and increasing market share. Helms brought in Max M. Bleck, previously with Cessna, as president and chief operating officer in 1978.[12] Sales exceeded $400 million for 1979, exports were strong and growing, and the future appeared favorable.

Frank Hedrick became president of Beech on January 18, 1968, as Olive Ann Beech, then age 65, assumed the office of chairman but remained active in management. Edward C. Burns, a nephew of the late Walter Beech, headed the Boulder Division and became regarded as heir apparent. Thus family influence continued, although family connections were never discussed publicly within the company. Howard "Pug" Piper, after his departure from Piper, joined Beech as a full-time consultant on light aircraft in 1973.[13] By that time possibly the strongest general aviation firm, Beech also became regarded as an attractive acquisition candidate, especially by larger aerospace firms. Merger discussions with Grumman, experiencing trouble in its naval business and seeking greater civil diversification, were broken off in October 1973 after Beech determined no mutually beneficial terms were possible.[14]

The Arab oil embargo in October 1973 was a major shock to Beech as well as to the entire general aviation sector. An initial government-imposed 42½ percent reduction in general aviation fuel allocations threatened to devastate the new aircraft market. The General Aviation Manufacturers Association led the effort to ease the restrictions. The reduction was modified to 25 percent and the market improved. In 1975 President Ford reduced general aviation export restrictions, also strengthening the market.[15]

The Musketeer series of smaller fixed-gear models was renamed in December 1971, with variants marketed as the Sundowner, Sport, and Sierra, and production continued of all three. The V-tail Bonanza and the Debonair, a derivative with a conventional tail, continued in production, and the Debonair was renamed the Model 33 Bonanza in 1967. The larger Model 36 six-seat straight-tail Bonanza appeared in mid-1968. The Duke, a new pressurized, piston-powered twin, sized between the Baron and the Queen Air, first flew on December 29, 1966, and entered production during 1968. Production of the Travel Air ended in 1968, but production and development of the popular Baron continued. Representing a considerable departure for Beech was the new Model 77 Skipper, a low-cost two-seat trainer with a raised cockpit and fixed tricycle gear. Intended for flying schools, the Skipper first flew on February 6, 1975.

At the higher end of the product line, production of the "Twin Beech" series ended in November 1969 after 32 years, then a record for a basic design. More than 750 of the definitive Super H18 variant had been built since 1964. Production of the Queen Air series began phasing out, ending finally in 1970. All larger business aircraft then were turbine powered. The initial King Air, effectively replacing the Super 18, was joined by the larger King Air 100, with more powerful PT6A-28 turboprops, in 1969. The lengthened, T-tailed Super King Air 200, flagship of the corporate line, made its first flight on October 27, 1972, and deliveries began in 1973. A new light twin intended for both business and multiengined training use, the Duchess 76, first flew on May 24, 1977, and attained major market share.

The first production Model 99, largest in the Beech line, flew on May 2, 1968, and was developed for both business and third-level airline or commuter use. The more powerful Model 99A appeared in 1969 and was joined by the Model 99 Executive, with a luxury corporate interior. The variants were later redesignated the B99 Airliner and B99 Executive. Sales of the B99 declined, however, and Beech suspended the line in 1975 with 164 produced, pending a market recovery.

One cloud over Beech, and over the general aviation industry, was that of liability for design-related crashes. The company became a target after four crashes of Beech aircraft were attributed to fuel cutoff during a "slip" turn and bank maneuver, causing power failure. Beech claimed pilot error but was accused of reluctance to face possible design flaws.[16] The company was not harmed at the time and went on to enjoy a prosperous decade in the 1970s, but the episode presaged a growing product liability problem.

While primarily a general aviation firm, Beech maintained its strong military involvement, including supersonic drones. Beech became a major subcontractor to the Bell JetRanger helicopter program in January 1968, producing airframes. Demands of that program on Wichita plant capacity were so great that Beech was forced to shift most general aviation production to the Salina plant. The company also built major subassemblies for the McDonnell Douglas F-4 fighter. The YT-34C turbine-powered model of the veteran trainer first flew on September 21, 1973, and 334 were delivered to the Navy between 1977 and 1984. All four military services ordered versions of the King Air for numerous missions, under various utility and transport designations.

Beech merger discussions in 1977 with General Dynamics, then interested in greater commercial diversification, were inconclusive.[17] Beech marked the delivery of the 10,000th V-tail Bonanza in February 1977 and in December of that year reached the production milestone of 40,000 aircraft since its founding in 1932.

Cessna retained its position as the largest general aviation manufacturer and offered the broadest product line. Unlike its competitors, it retained the high-wing configuration on all its single-engined models except for the agricultural line. In the military sector, the T-37 trainer was joined by the A-37B attack version in September 1967. The A-37B went into production for the U.S. forces and for export, a total of 577 being completed. Both the T-37 and A-37 remained in production until 1977. The Air Force ordered further T-41A trainers, and the Army ordered the more powerful T-41B. The type was also exported, and orders for military derivatives of the Model 172 totaled 855 by the end of 1978. The push-pull Model 337 was ordered by the Army for Vietnam service as the O-2A, for Forward Air Control (FAC), and as the O-2B for psychological warfare. Deliveries began in 1967 and extended into the mid-1970s.

Cessna reached the milestone of 10,000 Model 150s in December 1967, and in May 1972 became the first company to have produced 100,000 aircraft. Production of the Model 401 finally ended in 1972 after 406 were completed, but newer twin-engined executive models, including the Model 402, continued. More than 1000 of the

Model 402 were built, including convertible passenger/freight versions. Production of the Model 411A, which first appeared in 1968, ended in June 1978 with 303 having been delivered. The Models 414 and 414A Chancellor, appearing in 1969, represented the upscale, pressurized end of the market. A further refinement of the executive line was the Model 421 Golden Eagle, which first flew on October 14, 1965. Over 1500 Golden Eagles were produced.

While continuing the veteran Model 310, Cessna introduced the pressurized Model 340, an advanced development. Incorporating certain features of the larger 400 Series, the Model 340 appeared in December 1971 and effectively replaced the Model 320 Skyknight, which had ended production in 1968.

The Models 172 Skyhawk/175 Skylark and the Models 182/185/Skylane series continued their popularity throughout the 1970s, and further entries appeared in the twin-engined executive line. Departing from numerical designations, Cessna announced the Conquest, pressurized and powered by two TPE331 turboprops, on November 15, 1974, and the larger Titan, a piston-engined version, on July 16, 1975. Cessna's executive models held a competitive advantage in being priced considerably lower than the Beech King Air series. Cessna offered no fewer than 58 civil aircraft types in 1976.[18] The Model 152 two-seat model was introduced in 1977 to replace the veteran Model 150 and strong sales continued, but the market for the Model 337 Skymaster declined. In September 1967 Cessna introduced the Model 177 Cardinal luxury four-seat model, which found a strong market. It was joined in 1970 by the Cardinal RG with retractable gear, unusual in a high-wing design. More than 5,000 Cardinal and Cardinal RG models had been built when production ceased at the end of 1978 because of declining sales. Somewhat ironically, the Cardinal had been intended to replace the Skyhawk/Skylark series, but the older models remained in production.

The Pawnee Division produced single-engined models, while the Wallace Division produced military, twin-engined, and business jet models. The McCauley Division and fluid power activities also remained important. Although generally successful in all market segments, there was little question that the most important aspect of Cessna's business and the key to its future was the Citation business jet.

Russell W. Meyer, Jr., a lawyer and president of American Aviation Corporation until its acquisition by Grumman in 1973, was recruited by Dwane Wallace in June 1974 to become executive vice president of Cessna. Wallace, a strong believer in retirement at age 65, did so in 1976 after more than 40 years at Cessna. Meyer succeeded him as chairman and chief executive officer. Malcolm S. Harned, an industry veteran, became president.

Among the smaller competitors, Champion Aircraft Corporation, with an aircraft line descended from the original Aeronca, was acquired by Bellanca Sales Corporation on September 30, 1970, which was in turn the successor to the Inter-Air subsidiary of Miller Flying Service. After the Champion acquisition the firm was renamed Bellanca Aircraft Corporation. In addition to producing the Model 260C and Viking 300, Bellanca further developed the Citabria and also produced the Scout, the Champ, and the Decathlon, another development of the veteran Champion. The Champion operation, while very small, remained active in new model development, but eventually only the Citabria remained in production. Then the four-seat Bellanca Model 260C ended production in 1971. The Viking continued, and its largely wood structure marked a contrast with the dominant all-metal structures of competitors.

In 1976 Anderson, Greenwood, and Co., of Texas, inactive in aircraft production since the late 1940s, acquired 100 percent ownership of Bellanca Aircraft Corporation, still in Minnesota. The unrelated Bellanca Aircraft Engineering, Inc., headed by August Bellanca and descended from the original Bellanca company of 1927, was reorganized in 1971. Its advanced Skyrocket II five-seat business aircraft, under development for years, was first tested in 1975, but the company could not proceed with production.

A new Taylorcraft Aviation Corporation was formed on April 1, 1968, by members of the Taylor family, which held rights to the original designs. It provided parts and service support to the previous aircraft and from 1973 also built the new F-19 Sportsman, based on older Taylorcraft designs. Production of the F-19 ended in 1980, but the higher-powered F-21 continued in production at a low level.

Willard R. Custer, tireless proponent of his channel wing concept, retired in February 1968. He still pursued certification of the CCW-5 and promoted the aircraft, but his company was essentially dormant.

The Aerocar roadable airplane development of M. B. Taylor also largely ended by 1975. The combination of automotive safety, environmental, and fuel economy standards finally doomed market prospects. The model remained available by special order, however.

A new Navion Rangemaster Aircraft Company was organized in 1972, still in Texas, from the assets of the previous Navion Aircraft Corporation, which had gone bankrupt in 1969. The new company planned to reintroduce the Rangemaster into production from 1975. This venture also encountered market and financial problems, however, bringing to a close the many efforts to continue that classic light aircraft design.

Helio was acquired in 1969 by General Aircraft Corporation, also formed by Dr. Lynn L. Bollinger. It was renamed the Helio Aircraft Company. The large 10-seat Turbo-Stallion failed to sell in the civil market, and Helio effectively was forced out of business in the early 1970s. Production was suspended and the firm was sold in 1976, with the new owners moving to subcontracting.[19] Rights to the Helio designs were again sold, but no new production resulted.

Swearingen progressed with production of both the piston-powered Merlin I and the turboprop-powered Merlin II. From these were developed the uprated and refined Merlin III, followed by the further upgraded Merlin IIIB, which first flew on December 1, 1968. Even more ambitious was the SA-226A Merlin IV, which involved a major stretch of the fuselage. The plane first flew on August 26, 1969, and was intended for both corporate and commuter use. The commuter version, renamed Metro, was developed in a joint venture with Fairchild Hiller. The Metro first flew on August 29, 1969, and the type soon captured a share of the 19-seat third-level airliner market segment. After Swearingen experienced financial difficulties, Fairchild acquired 90 percent ownership on November 2, 1971, operating it as the Swearingen Aviation Corporation subsidiary. It was still managed by Ed Swearingen. Production of earlier models ceased, but Fairchild marketed the improved Merlin IVA as the Metro II for the third-level segment. Fairchild acquired the remaining 10 percent of the stock in 1979.

Maule Aircraft Corporation continued with its M-4 Rocket series, which gained a reputation for ruggedness. The model was also produced in Mexico from 1964 as the Cuauhtemoc M-1, although only for the Mexican Air Force. By 1968 some 250 aircraft had been delivered, including the export market, and Belford D. Maule began considering a new location that would offer lower labor costs, better flying weather, and expansion space. He selected a closed air base in Moultrie, Georgia, and moved operations there in September 1968. Production of the Maule M-4 ended in 1975, but the M-5 Lunar Rocket, which first flew on November 1, 1971, and subsequent developments enjoyed long production lives.

The Lake Aircraft Division of Consolidated Aeronautics continued as the only producer of small amphibians. The basic LA-4 model continued to be refined. The firm was reconstituted in 1981 as Lake Amphibians, Inc., of Laconia, New Hampshire, under control of the French entrepreneur Armand Rivard.

A rare success in a new general aviation venture was the American Aviation Corporation, formed in 1964 in Cleveland, Ohio, with Russell W. Meyer, Jr. as president. American Aviation had acquired the rights to the BD-1 design of James Bede and his Bede Aviation Corporation, which it developed as AA-1 Yankee. The small Yankee gained attention for its low-cost aluminum honeycomb construction, as well as for its STOL airfield performance. By 1972, it achieved a measure of sales success in the two-seat segment, long dominated by the Cessna 150.

Grumman, implementing its strategy of reducing its dependence on the U.S. Navy, acquired American Aviation Corporation on January 2, 1973. The subsidiary was renamed Grumman American Aviation Corporation (GAAC), but operations remained in Cleveland. President Russ Meyer was succeeded in May 1973 by the unrelated Corwin (Corky) Meyer, a Grumman vice president with a long background as a test pilot. Grumman's Ag-Cat, still built under license by Schweizer, and Gulfstream II models then were brought under GAAC, giving Grumman a broad range of general aviation aircraft. The turboprop Gulfstream I ended production in February 1969 with 200 delivered. More than 2,000 Ag-Cats had been produced by 1979.

With Grumman financial support, GAAC introduced new models from 1974, including the Trainer and the AA-5 Traveler four-seat model, both from the AA-1 Yankee, and the Model Tr 2 combined trainer and sportplane. The Grumman tradition of feline names continued with the advanced AA-5B Tiger. Then the GA-7 Cougar, a completely new light twin, first flew on December 20, 1974. GAAC also seriously studied during 1979–1980 a 30–40 seat commuter airliner development of the Gulfstream I, but decided not to proceed. (It appeared logical to many for larger general aviation firms to move into 30–50 seat commuter airliners. NASA, in fact, sponsored a Small Transport Aircraft Technology (STAT) program to advance commuter aircraft technology.[20] No firm produced a new design, however, primarily owing to a poor record of profitability in that segment. That category was left to foreign producers, many with government subsidies.)

Grumman's increased investment in general aviation proved to be of short duration, however. In another strategic move it sold its GAAC interests to the aviation entrepreneur Allen Paulson through his American Jet Industries of California. American Jet Industries (AJI) purchased Grumman's 80 percent shareholding on September 1, 1978, and purchased the remaining 20 percent held by others with cash. AJI, which had earlier acquired the original Fletcher company, had determined to reenter aircraft production. In 1970 it purchased the new plant built by Lockheed to produce its Cheyenne attack helicopter, which was cancelled, and centered its activities there. After purchasing the Grumman interests, Paulson combined all operations as Gulfstream American Corporation and headquarters moved to the Savannah factory. Gulfstream, however, soon ended production of the small single-engined designs, then of the Cougar in 1980.

Helicopter Survey

The established general aviation helicopter producers experienced frequent ownership changes during this period. Financial struggles reflected the difficulty of a small civil helicopter market more than any other factor. Helicopter usage expanded for hospitals and police forces, however, where its capabilities could not be matched by fixed-wing aircraft. In addition, business and executive use of helicopters finally began to grow in the late 1970s. Turbine-powered smaller models became more commonplace, and the Allison 250 turboshaft series became an industry standard. A difference between fixed-wing and helicopter marketing was that helicopter sales generally were negotiated directly from the factory; dealerships had never been widely established. A narrower market and fleet orders were the major reasons. Exports became increasingly significant (see Table 6-4).

Brantly, previously acquired by Lear and included in the Lear acquisition by Gates, was sold early in 1969 to the new Aeronautical Research and Development Corporation (ARDC), led by the noted defense attorney and helicopter enthusiast F. Lee Bailey. Late in 1970, ARDC was in turn acquired by Brantly Operators, Inc., of Lakeland, Florida, led by Michael K. Hynes. Hynes briefly considered but rejected acquiring the FH-1100 from Fairchild in 1973. The company became Brantly-Hynes Helicopter, Inc., on January 1, 1975, reflecting the ownership interest of Hynes, and operations were moved to Frederick, Oklahoma. The Model B-2B and Model 305 remained in production.

Enstrom was acquired by Purex Corporation in 1968, becoming a subsidiary of its Pacific Airmotive (PacAero) operation. Poor sales led PacAero to shut down production in February 1970, however. Enstrom then was acquired in January 1971 by an investor group led by F. Lee Bailey. Having just sold Brantly, Bailey undertook development and marketing efforts in his second helicopter venture. By 1977 500 Enstrom helicopters had been delivered, and the turbine-powered Model 480 had appeared.

A new contender in the small personal helicopter market appearing in the 1970s was the Robinson Helicopter Company of Torrance, California. Franklin Robinson, an experienced helicopter engineer, began designing the two-seat R22 in June 1973, and the prototype made its first flight on August 28, 1975. First deliveries were in October 1979, and output expanded gradually. Emphasizing low cost, and powered by a 160 hp Lycoming piston engine, the R22 soon led the light two-seat helicopter segment.

Fairchild Hiller transferred its remaining helicopter production to Hagerstown in 1966 in order to centralize all light aircraft production. Production of the veteran Model 12 ended in 1967. Although market prospects for the FH-1100 initially appeared favorable, rising prices and increasing insurance costs diminished the market. Competition from the Bell JetRanger was also a factor, and FH-1100 production ended in 1971 with 250 sold. Fairchild Industries, Inc., renamed from Fairchild Hiller Corporation in 1971, disposed of the Hiller helicopter operations in its restructuring of January 1973. In the same month a new Hiller Aviation Company was formed by

Table 6-4
Exports of Commercial Helicopters by U.S. Manufacturers, 1960–1987

Year	Total output	Export number*	Export $ value (000's)
1960	266	89	$ 11,446
1961	744	122	10,483
1962	407	78	11,124
1963	504	60	10,982
1964	579	102	20,080
1965	598	173	25,121
1966	583	161	11,500
1967	455**	223	25,200
1968	522**	242	32,900
1969	534	252	29,100
1970	482	332	26,900
1971	469	298	45,700
1972	575	256	50,300
1973	770	428	83,300
1974	828	395	107,600
1975	864	336	104,700
1976	757	315	113,000
1977	848	321	106,000
1978	904	368	156,000
1979	1,019	459	207,000
1980	1,366	525	249,000
1981	1,072	453	346,000
1982	587	254	206,000
1983	401	216	232,000
1984	376	233	234,000
1985	376	137	210,000
1986	326	210	277,000
1987	358	281	239,000

*Export figures include Bell, Fairchild Hiller, Hughes, Sikorsky, and Vertol.

**Excludes foreign licensees of Bell.

SOURCE: *Aerospace Facts and Figures, 1969,* pp. 31 and 37; *Aerospace Facts and Figures, 1977–1978,* p. 112; *Aerospace Facts and Figures, 1981–1982,* p. 35; *Aerospace Facts and Figures, 1988–1989,* pp. 131–132.

Stanley Hiller in California. It acquired all helicopter rights with the exception of the FH-1100 from Fairchild. Hiller Aviation resumed production of the UH-12E for the civil market.

The light four-seat Bell Model 206 JetRanger and at the high end, the twelve-seat twin-turboshaft Sikorsky S-76, became increasingly popular executive transports. The S-76 first flew on March 13, 1977, and entered service early in 1979. Sikorsky, hitherto almost exclusively military, regarded the S-76 as a means of increasing civil market share. The Bell JetRanger led to a long series of civil developments. The initial Model 206A was succeeded in 1972 by the Model 206B JetRanger II, with a more powerful Allison 250, followed by the further improved JetRanger III from 1977. The larger Model 206L LongRanger, announced in September 1973, filled the gap between the JetRanger and heavier types. More than 5,000 Model 206s had been built by 1977, including some 2,200 civil models. The Model 47, the original general aviation helicopter, finally ended production in 1976.

Bell's large Model 212, based on the military UH-1N and marketed as the Twin Two-Twelve, appeared in 1971. The Model 412, an update of the Model 212, followed in August 1979. Then the new intermediate Model 222, the first commercial twin-turbine helicopter built in the United States, first flew on August 13, 1976, and became popular both for executive use and for servicing oil drilling platforms.

Hughes Helicopters remained a minor presence with its Model 300, competitive with the JetRanger.

Business Jet Progress

The corporate jet segment, represented in the United States by the JetStar, Sabreliner, Gulfstream II, the LearJet series, and until 1968 the Jet Commander, became increasingly important to general aviation. While initial sales were slow, the market expanded rapidly in 1965 and 1966. LearJet sales totaled 305 by the end of 1970, making it by far the most popular of the business jets. Bill Lear departed on March 31, 1969, after his sale to Gates, and the company became Gates Learjet Corporation in January 1970, with production still in Wichita. Lear pursued further business aviation ventures.

Although hurt by the recession of 1970 and 1971, the executive jet market soon experienced a strong recovery, led by the LearJet, Gulfstream II, and the French Falcon. The market in fact expanded into three distinct segments: the small, represented by the LearJet and the new Cessna Citation; the medium, represented by the Sabreliner, the Falcon 20, and the British BAe 125 (originally DH-125); and the large, represented by Gulfstream II and later the Canadair Challenger. The market also became strongly international, as American designs competed with those from Britain, France, and Israel. Pan Am established a Business Jets Division to market the Dassault Falcon in North America. Israeli Aircraft Industries (IAI) undertook a long series of developments of the original Jet Commander acquired from North American Rockwell, resulting first in the IAI Westwind and ultimately in the Astra Jet. The Canadair Challenger, entering the market in 1982, and the Mitsubishi Diamond also became factors. Mitsubishi, expanding its Texas subsidiary where the MU-2 remained in production, brought the locally developed twin-jet MU-300 Diamond, which first flew on August 29, 1978, to the U.S. market.

Cessna's business jet strategy was to offer a smaller model for the lower-cost segment. It announced the straight-winged Fanjet 500, powered by twin CJ610 turbofans, in 1968. Soon renamed the Citation, it first flew on September 15, 1969. But with the business jet field becoming overcrowded and with a development expenditure in the $20 million range, the Citation represented a major financial risk.[21] Furthermore, initial sales of the Citation were sluggish when it entered the market in 1972. Cessna maintained, however, that its simpler and somewhat slower design would motivate operators of turboprop-engined business aircraft such as the King Air to move up to a business jet. Sales eventually rose, and the Citation series became numerically the most prolific of all business jets. Numerical suffixes began with the larger Citation II, announced in September 1976, and the improved basic Citation I, which followed in December 1976. The Citation II first flew on January 31, 1977.

Beech, unlike Cessna, Lear Jet, Grumman, and Aero Commander, did not develop an original business jet. Instead, it entered into a marketing and distribution agreement with Hawker Siddeley (later British Aerospace) to form Beechcraft Hawker Corporation in 1970, marketing the originally de Havilland DH-125 in North America as the Beechcraft Hawker BH-125. But the Hawker marketing agreement was ended by Beech in 1975, the reason being simple economics.[22]

Grumman had planned a refined Gulfstream III, but development was halted in 1977 in favor of a later, even more advanced version incorporating a highly efficient NASA-developed supercritical wing and winglets. Having sold the Gulfstream to Gulfstream American, Grumman resumed development in 1978 under contract to the new owner, and the new Gulfstream III first flew on December 2, 1979. Grumman continued as a production subcontractor to Gulfstream as well. Gulfstream II production ended in 1980 with delivery of 256, succeeded immediately by the Gulfstream III.

Lockheed introduced the developed JetStar II, with advanced Garrett TFE 731-3 turbofans, in the summer of 1976. But after production of 40, the JetStar program ended in 1979. More than 200 of all versions had been produced. Rockwell International continued to develop and manufacture the Sabreliner in Los Angeles, although marketing was controlled by a new division in St. Louis. All Sabreliner production and support operations then were transferred to St. Louis in March 1977, by which time some 500 had been delivered. Both the Model 60 and larger Model 75 remained in production.

Gates Learjet announced major model improvements on October 28, 1975, including the advanced, long-range Model 55. It followed with the Models 28/29 Longhorn, improvements of the Model 25, featuring longer wings and winglets. First flight came on August 24, 1977. Further developments included the Models 35A/36A, also similar to the Model 25, but with turbofan power. Airframe production remained in Wichita, but "green" aircraft were transferred to a facility in Tucson, Arizona, for final furnishing and equipment.

The business jet market continued its generally strong growth through the 1970s. Production stability was aided by recurring military orders for certain models for training and high-priority transport missions. Governments also ordered business jets for transportation of high officials. Business jets were marketed on the premise that they would save valuable management time over scheduled commercial air travel, thereby maximizing productivity as well as bringing remote plants and other operations not served directly by commercial airlines within easy reach. The increasing popularity of hub-and-spoke routing by airlines, causing inconveniences for many business travelers, was a further incentive.

Table 6-5
General Aviation Shipments, 1976–1980

Firm	1976	1977	1978	1979	1980
Ayres	—	—	134	99	44
Beech	1,220	1,203	1,367	1,508	1,394
Bellanca	315	252	370	443	103
Cessna	7,888	8,839	8,770	8,400	6,393
Gates Learjet	84	105	102	107	120
Gulfstream	762	866	933	400	167
Lake	88	99	98	96	79
Lockheed Jetstar	3	16	9	7	4
Maule	96	108	88	67	59
Mooney	227	362	379	439	332
Piper	4,042	4,499	5,272	5,255	2,954
Rockwell	595	432	244	164	146
Swearingen	30	28	51	70	86
Ted Smith Aerostar	100	101	—	—	—
Totals	15,450	16,910	17,817	17,055	11,881

SOURCE: *Aerospace Facts and Figures, 1981–1982,* p. 35; and GAMA reports.

The productivity case was supported with statistics showing that corporations owning and operating business jets enjoyed higher profitability than those which did not. The objective was to establish causality between business jet use and profitability, although that remained tenuous. More realistically, given their high costs, the economics of the corporate jet for most companies appeared rather dubious. Further, not all successful firms used business jets. Boeing, for example, never owned or operated corporate jets, relying entirely on commercial airlines for its travel needs. Instead, ownership of corporate jets and entitlement to their use became known as the ultimate corporate perk.[23]

With a rise in sabotage and terrorism affecting commercial airlines, corporate jets increasingly were marketed on security grounds. Then as the market matured, many jets owned by corporations were chartered or leased during periods of inactivity. Manufacturers also offered leasing to their corporate customers as an alternative to the relentless escalation in selling prices, as well as accepting older models in trade toward new corporate jets. Specialized business jet leasing firms also appeared.

# # # # #

Despite economic fluctuations and periodic political disturbances, general aviation manufacturing had enjoyed a strongly expanding market and wide prosperity during the 1970s (see Table 6-5). That strength and prosperity were not to continue, however. By the beginning of the new decade, several factors converged to nearly devastate the industry.

7

The General Aviation Crisis

General aviation manufacturers appeared well positioned to continue their strong growth into the 1980s. Production volume, dollar sales, and employment peaked in 1979, and optimism for the future was widespread. Cessna led with sales of $939 million for that year, followed by Beech with $602 million and Gates Learjet with $302 million, all including military sales.[1] Total general aviation aircraft sales, $2 billion in 1979, were expected to reach $3 billion in 1981.* But as the American economy enjoyed a long overall expansion during the decade of the 1980s, the general aviation industry declined sharply, and the decline turned out to be severe and long term rather than cyclical. The reasons were many, but the most damaging factor was the rash of product liability lawsuits against manufacturers stemming from aircraft accidents.

Prices had already escalated substantially in the 1970s. Liability insurance costs forced manufacturers to raise prices even more, sharply diminishing new aircraft demand. The market decline was felt earliest and most sharply in the single-engined segment, as growth continued until 1981 in twin-engined propeller-driven models and business jets. A record 389 business jets and 918 turboprop-powered models were delivered in 1981, but overall executive aircraft production declined sharply thereafter. Exports of general aviation aircraft (see Table 7-1), which generally had ranged from 20 percent to 35 percent of total output, accounted for more than 40 percent by 1990 due to the domestic market collapse.[2] Concurrently, imports of general aviation aircraft into the U.S. market increased sharply, reaching 35 percent by value in 1985, although concentrated in the high-value business jet and turbine-powered commuter segments rather than in small single-engined types.[3]

* The sales figures represented fixed-winged aircraft only, as rotary-winged aircraft were statistically separate.

Yet U.S. production still accounted for approximately 70 percent of all turbine-powered general aviation aircraft in operation, and the United States remained the largest general aviation market in the world. It was estimated in 1984 that some 36,000 U.S. businesses used aircraft. Product development remained active, particularly at the high end of the market, even with sales declining (see Table 7-2). In the midst of the market decline, the general aviation community noted the passing in 1982 of Matty Laird, a pioneer of the industry, at age 87.

Table 7-1
General Aviation Exports, 1959–1995

Year	Total units	Helicopter exports	GA exports
1959	1,033	(not reported until 1969)	
1960	1,528		
1961	1,646		
1962	1,458		
1963	1,583		
1964	1,834		
1965	2,457		
1966	2,985		
1967	3,125		
1968	2,879		
1969	2,713	252	2,461
1970	2,369	332	2,037
1971	1,864	298	1,566
1972	2,328	256	2,072
1973	3,591	428	3,163
1974	4,658	395	4,263
1975	3,604	336	3,268
1976	3,533	315	3,218
1977	3,790	321	3,469
1978	4,839	368	4,471
1979	4,337	459	3,878
1980	3,703	525	3,178
1981	3,070	453	2,617
1982	1,194	254	940
1983	735	216	519

Year	Total units	Helicopter exports	GA exports
1984	658	233	425
1985	621	137	484
1986	674	210	464
1987	729	242	487
1988	923	280	643
1989	1,604	294	1,310
1990	1,158	349	809
1991	852	318	534
1992	570	212	358
1993	508	175	333
1994	539	154	385
1995	573	210	363

SOURCE: *Aerospace Facts and Figures, 1969,* p. 73; and *Aerospace Facts and Figures 1996/1997,* p. 30, based on company reports, GAMA, International Trade Administration, Department of Commerce.

Table 7-2
Annual Production and Sales, General Aviation Aircraft

Year	Units	Factory sales ($ millions)
1981	9,457	2,919.9
1982	4,226	1,999.5
1983	2,691	1,469.5
1984	2,431	1,680.7
1985	2,029	1,430.6
1986	1,495	1,261.9
1987	1,085	1,363.5
1988	1,143	1,918.4
1989	1,535	1,803.9
1990	1,144	2,007.5
1991	1,021	1,968.3
1992	899	1,839.6
1993	964	2,143.8
1994	928	2,357.1
1995	1,077	2,841.9
1996	1,132	3,141.0

SOURCE: *General Aviation Statistical Databook (1996 ed.),* p. 4.

Industry Survey

Piper's cumulative production reached 125,000 on May 4, 1982, but by that time output was falling rapidly with the market decline. After J. Lynn Helms departed for the FAA in 1981, Max M. Bleck, already chief executive officer, succeeded him as chairman as well. Piper then acquired a new corporate parent when Lear Siegler Corporation acquired Bangor Punta on March 1, 1984, for $282 million.

Continuing attempts to broaden its market coverage, Piper formed an Airline Division on June 4, 1981, to support third-level operators of the Chieftain and to produce new versions for that market. The piston-engined T-1020 and the turboprop-powered T-1040, both roomier developments of the Chieftain, were offered. The Airline Division became inactive in 1986, however, after only 23 T-1020 and 23 T-1040 models had been produced.

Aerostar operations, under Piper ownership, were transferred to Vero Beach in October 1981. Several variants of the basic Aerostar, differing in pressurization, power, and equipment, were offered. But production of the Aerostar 601P was suspended in 1982 due to weak market conditions, followed by suspension of the Model 600A. Total Aerostar production of Models 600, 601B, and 601P by all corporate owners totaled 875. Then the remaining models 602P and 700P were suspended in 1984, bringing the production history of that design to an end.

Production of the PA-38 Tomahawk, PA-23-250 Aztec F, and the PA-44 Seminole light twin, which had entered service only in 1978, was suspended in 1982. Rights for the veteran PA-18 Super Cub and the PA-36 Brave were sold to WTA, Inc., of Lubbock, Texas, in 1983, but both models were still manufactured by Piper on behalf of WTA. Then the Super Cub operation ceased in 1987. Piper Navajo production also ended, but product development continued with the entirely new PA-46 Malibu, a large single-engined business aircraft offering twin-engined cabin comfort. Announced on November 20, 1982, the Malibu supplemented the slow-selling Saratoga, and was regarded as the key to Piper's future in the declining market.

Beech finally gave up its independence on October 1, 1979, with its acquisition by the Raytheon Corporation of Lexington, Massachusetts. The culmination of years of merger considerations, Beech agreed to the friendly takeover after being assured that it would retain its distinct identity, a concern which had weighed against any agreement with Grumman or General Dynamics in earlier talks. The acquisition also promised synergy: there were no overlapping product lines; Raytheon would gain commercial diversification; and Beech would benefit from Raytheon's technical capabilities, particularly in defense contracts.[4] It was anticipated that Beech would become the largest profit contributor to Raytheon. Another concern was that of aging management: Mrs. Beech was 76 at the time of the merger and Frank Hedrick was 69.

The merger was approved by stockholders on February 6, 1980, and became effective on February 8.[5] Payment was by an exchange of stock, with the transaction valued at some $800 million, an advantageous price to Beech but one which later brought Raytheon intense criticism, coming as it did just before the market decline.[6] Olive Ann Beech and Frank Hedrick became directors of Raytheon. A series of management changes followed as Beech nephew Edward C. Burns became president on January 10, 1981, with Frank Hedrick moving to vice chairman. Burns was forced to make significant layoffs during 1982, when Beech, along with other firms, was adversely affected by the market collapse. Burns then retired on June 15, 1982, followed by Hedrick on July 1, as Raytheon management influence increased. Olive Ann Beech retired as chairman of the board of Beech in September 1982, acquiring the title of chairman emeritus, and D. Brainerd Holmes, chairman of Raytheon, also became chairman of Beech. The aviation entrepreneur Linden S. Blue, formerly with Learjet and Lear Fan, Inc., succeeded Burns as president and CEO but served only 26 months before departing. He was followed in those positions by James C. Walsh in 1984. Then Max M. Bleck, formerly CEO of Piper and later with Learjet, became president and CEO of Beech in 1987.

Beech entered business jet production in December 1985 by purchasing rights to the developed Mitsubishi Diamond 2, which had first flown on January 28, 1985. Mitsubishi, which had ended its cooperative agreement with Mooney, also decided not to continue Diamond 2 production in San Angelo and built only 11 before the sale to Beech. More than 500 MU-2 models had been built, but Mitsubishi closed all its general aviation activities on March 31, 1986, and exited the market. Beech took over production of the renamed

Beechjet, still using Mitsubishi components, but total transfer of manufacturing to Wichita was completed in June 1989. Demand developed slowly, chiefly due to competition from the established Citation and Learjet, but the improved Beechjet 400A, manufactured entirely in Wichita, boosted sales.

A revised Beech Model C99, a 15-passenger model for third-level service developed from earlier versions, was announced on May 7, 1979, and first flew on June 20, 1980. It was placed in production at a new facility in Selma, Alabama, on the site of the closed Craig Air Force Base. Then the successor to the Model 99 series, the completely new Beech 1900C, first flew on September 3, 1982, and entered production in 1983. Possessing 40 percent design and component commonality with the Super King Air 200, the 1900C could accommodate 19 passengers.

Beech continued to rationalize its product line, even as it developed new and improved models. Production of the Duke, the King Air 100, and the Duchess 76 ended in 1984. The E55 Baron ceased production after 1,201 had been completed, but the improved E58 Baron continued. Beech also suspended production of the single-engined Sierra and Sundowner in 1984. The Model V35 Bonanza, finally bowing to diminished demand and increasing questions over the safety of its V-tail design, ended its 38-year production life early in 1985, with 10,390 built over 38 years. Beech had consistently denied recurring accusations that it covered up safety problems with the design. Straight-tail Bonanza production continued, as did the twin-turboprop line. The Super King Air 300 was succeeded by the much-improved Super King Air 350 late in 1989. The King Air C90A and Super King Air 200 continued concurrently with the more advanced model, and military versions continued to receive orders.

Criticized at times for its design conservatism, Beech replied by designing the radical Starship 2000 beginning in 1982. Featuring an innovative canard configuration, twin pusher turboprop power, and a structure of advanced composite materials, the Starship was also to be competitive in speed with business jets. Projected as the eventual replacement for the King Air series, but with a development cost of some $350 million, the Starship represented a major risk for Raytheon. A proof-of-concept test model at 85 percent scale, built by Burt Rutan's Scaled Composites development firm in California, was first flown on August 29, 1983. Then the full-scale prototype was first flown on February 15, 1986. Scaled Composites, Inc., founded in 1982, was acquired by Beech in June 1985, but was then resold to Burt Rutan in November 1988, reportedly after some differences between Rutan and Beech management.

Cessna, successful with smaller business jets, moved up the scale with the larger Citation III, which first flew on May 30, 1979. With a supercritical swept wing, the Citation III went into production in 1980. Cessna delivered the 1000th Citation early in 1982. The Citation S/II, with a supercritical wing, appeared in 1984, succeeding the earlier straight-winged models. The lower-cost Citation VI and more powerful Citation VII were developments of the Citation III, but the Citation IV, also developed from the Citation III, was canceled in 1990.

Another major product development was the Caravan, a capacious single-turboprop utility design which first flew on December 9, 1982. A logical step up for Cessna from its Skywagon and Stationair utility models, as well as a potential replacement for the veteran Canadian Beaver and Otter bush aircraft, the Caravan could also be equipped with floats. The Caravan captured a substantial worldwide market, and in addition was ordered by the Air Force as the U-27A for supply to foreign air arms under the Foreign Military Sales (FMS) program.

Cessna's cumulative production total reached 172,000 in 1982, but production of smaller models declined rapidly. The push-pull Model 337 ended in April 1980, and the veteran Models 180 and 310 ended their long production lives in 1981. More than 5,000 of the 310 series had been built. A new T303 Crusader light twin, initially powered by 160 hp piston engines, first flew on February 14, 1978. Entering production and service in 1981 with 250 hp engines, the T303 was suspended in 1985 in the continuing retrenchment by Cessna. In any event the T303, while more spacious, became generally regarded as somewhat inferior to the 310 series. The three Cessna aircraft production plants were consolidated into a single aircraft division in 1984, reflecting the drop in demand, and the move also involved major layoffs.

The last major independent general aviation firm, Cessna agreed to become a subsidiary of General Dynamics Corporation on March 3, 1985, for $660 million cash. General Dynamics chairman David S.

Lewis already served on the Cessna board. The merger became effective on September 13, 1985. Dwane Wallace, possibly the most consistently successful executive in the history of general aviation, died in 1989, aged 77. He reportedly had become disenchanted with his successor, Russ Meyer, and left the board and sold most of his Cessna stock in 1983.

Remaining smaller general aviation firms sustained production generally at sharply reduced levels. Republic Steel, owner of Mooney, was acquired by LTV Corporation in 1984. Continuing its ownership odyssey, Mooney thereupon was sold to a private investor group which formed the Mooney Holding Company. The subsidiary Mooney Aircraft Corporation was under the management of LeRoy P. LoPresti, previously with Grumman American. Then French investors, led by the aviation entrepreneur Alexandre Couvelaire, acquired 70 percent of Mooney in the spring of 1985. The remaining 30 percent was held by Armand Rivard, who also controlled Lake Amphibians. A restructuring in February 1986 was followed by Rivard's sale of his interest in September. Couvelaire served as chairman of Mooney and controlled strategy.[7] Robert Cromer was brought in as general manager, and Jacques Esculier later became CEO under Couvelaire. Annual production fell from more than 400 in 1979 to 90 in 1985, but development continued of the basic four-seat low-wing retractable-gear M-20 design. Al Mooney, long retired from Lockheed, died in 1985 at age 79.

Production of the Lake LA-4 Buccaneer continued steadily in New Hampshire, with the 1000th model of the series delivered in 1980. The Buccaneer was phased out of production in 1984, being succeeded by the refined Renegade, which could seat up to six. The factory relocated to Kissimmee, Florida, in 1987, although headquarters remained in Laconia. The Renegade, while rather expensive for the time at some $250,000, remained the only piston-powered amphibian in production in the world. The market was, in fact, primarily international for the amphibian. The Renegade was joined by the Turbo 270 Renegade and by the SeaFury model for saltwater operation.

Bellanca production consisted of the Viking and the new Aries T-250, an Anderson, Greenwood design. But with increasing financial problems due to declining business, Anderson, Greenwood, and Co., liquidated the Bellanca Division in 1981. Then Viking Aviation, organized by most of the former managers and officers of Bellanca, purchased the rights and organized a new Bellanca, Inc., on May 7, 1982. Production of the Viking eventually was reinstated in 1984 at low volume. The firm filled an order from the Turkish Army for 30 Citabria trainers during 1982–1983, the only military order for the type. The Champion line was sold by Bellanca, Inc., in August 1982 to a new Champion Aircraft Company in Houston, Texas, but that venture failed in 1985.

Ayres maintained low-volume production of its Thrush Commander line of agricultural aircraft. But as was the case with other market segments, demand for agricultural aircraft declined. Maule Aircraft Corporation entered Chapter 11 bankruptcy late in 1984, largely as a defense against product liability suits, but the successor, Maule Air, Inc., resumed production of the upgraded MX-7 series from its veteran basic design. Taylorcraft, with a low level of activity, was purchased by a group of former Piper executives, including William T. Piper, Jr., on July 9, 1985. Production was moved to the recently closed Piper factory in Lock Haven, but the company soon became financially shaky again and entered Chapter 11 proceedings late in 1986.

Fairchild's Swearingen subsidiary, exclusively in the 19-seat commuter segment, was renamed Fairchild Aircraft Corporation in September 1982 but remained operationally autonomous from the parent Fairchild Industries. Edward J. Swearingen departed Fairchild Aircraft afterward and established a new Swearingen Aircraft Corporation at the end of the year to pursue new business aircraft developments.

Spurred in part by its difficulties in the military and commercial sectors at the time, Fairchild Industries sold Fairchild Aircraft Corporation and rights to the Metro to GMF Investments of California, a venture capital firm. The transaction took place in December 1987, after negotiations with Aeritalia of Italy fell through. The sale appeared to end the uncertainty over the Metro, and the improved Metro III, with longer wingspan and four-blade propellers, appeared in 1988. But under its new owners, the company soon became embroiled in a lawsuit with British Aerospace Corporation in which BAe alleged that Fairchild had acted to undermine prospects in North America for its competitive Jetstream. Fairchild first sued BAe late in 1988, but BAe responded with a $225 million countersuit.

After Sanwa, its major Japanese financier, withheld credit, Fairchild Aircraft declared Chapter 11 bankruptcy on February 1, 1990, to defend itself against the contingent liability created by the lawsuit. GMF ousted the Fairchild management but continued production at low volume.[8] There were subsequent discussions with Mooney's French owners of a Fairchild-Mooney combination, but in September 1990 the firm was acquired by Carl Albert, formerly a regional airline executive, backed by $35 million in new Japanese financing. Metro production continued, including the C-26 version for the Army.[9]

Gulfstream American continued to expand in general aviation, and its success with the Gulfstream made it a major firm along with Beech, Cessna, Piper, and Learjet. Rockwell International had reluctantly decided that the single-engined and business aircraft market was not viable for a primarily military firm. Gulfstream purchased the RI General Aviation Division on February 3, 1981, effectively ending Rockwell's involvement but enhancing Gulfstream's stature in general aviation. Its status as a comprehensive general aviation manufacturer would also prove to be of short duration, however, as its Commander Division soon ended production of the Shrike Commander. Production continued of the pressurized Aero Commander models 840, 900, 980, and 1000. The models 840 and 980 originally were intended to replace the Turbo Commander 690B, but production of those models also ceased late in 1983. Only 21 Turbo Commander models were delivered in 1984, reflecting to a high degree the seriously overcrowded twin-engined business aircraft segment.

The market decline then led to the phaseout of all original Aero Commander models, with production permanently discontinued in January 1985. The Bethany, Oklahoma, production facility was closed.[10] That event marked the end of more than 35 years of production, extending from the late 1940s and totaling some 3,000 units, of the basic twin-engined Aero Commander design.

Having ended production of the former Grumman American models earlier, Gulfstream American again became an essentially single-product company. It initiated the extensively developed Gulfstream IV in April 1982, with a lengthened fuselage, structurally redesigned wings, and more advanced Rolls-Royce Tay engines. The Gulfstream IV first flew on September 19, 1985, and gained immediate market acceptance. Although costing in the $25 million range, the Gulfstream IV held a market advantage in that potential purchasers were less likely to be affected by a recession than those considering smaller jets. Grumman remained a subcontractor, and wings were manufactured by Vought Aircraft in Texas.

Allen Paulson renamed the firm Gulfstream Aerospace Corporation on November 15, 1982. He sold the company to Chrysler Corporation on August 16, 1985, for $637 million, but remained as manager. Chrysler, strongly profitable at the time but the least diversified of the auto companies, viewed aerospace diversification as a priority. In a later automobile market downturn, however, Chrysler decided that it could not afford the investment necessary to keep Gulfstream competitive. Paulson then repurchased the company from Chrysler, through the investment firm Forstmann Little, in February 1989, paying a premium price of $825 million.[11] Forstmann Little became the major stockholder in Gulfstream.

Rockwell International disposed of its last civil aircraft program, the Sabreliner, by selling the Sabreliner Corporation subsidiary to the investment banking firm of Wolsey and Co. in July 1983. The operation remained in St. Louis, but was active primarily in modification and product support. Sabreliner development and production nevertheless continued at low volume.

Production of the Learjet Models 28/29 Longhorn was suspended in 1982 due to the slow market. The Model 55 Longhorn, a major redesign and the first of the larger 50-series, first flew on April 19, 1979, and entered production. The Longhorn name was later dropped. By the end of 1984 1000 Learjets had been produced, but all production was suspended in that year due to market conditions. In addition, the radical Lear Fan 2100 pusher-engined development, undertaken by a separate company with Saudi financial backing, was terminated in June 1985, with FAA certification unattained after some $120 million had been spent.[12]

When Learjet production resumed, the Model 31, combining the fuselage of the Model 35/36 and the new wing of the Model 55, was delivered from 1987. The Air Force ordered 80 Model 35s as the C-21A. Gates Learjet, holding a valuable tax loss carryback, in September 1987 sold a 64.8 percent interest to Integrated Acquisition, Inc., a subsidiary of Integrated Resources, primarily a real estate investment firm. The firm was renamed Learjet Corporation, and production returned entirely to Wichita by January 1988.

The Homebuilt Sector Grows

Development of a viable homebuilt or kitplane segment of general aviation had been largely achieved by the late 1960s. The major incentive remained that steadily escalating prices of factory-built aircraft enhanced the cost savings of kit construction, a primary consideration of sport and recreational pilots. Pride of construction also remained a factor, but a major obstacle remained in that the market for very small single-seat sportplanes, many of which were biplanes, was very narrow. Thus development of a stable and profitable business on that basis appeared almost unattainable. Those interested retained their enthusiasm, but there appeared to be little potential for a substantial industry in that category. Enthusiasts for such aircraft almost inevitably built them for personal use only.

Another obstacle to development of the homebuilt segment was that all required certification by the FAA in either the normal or utility categories, which often was protracted. But recognition of the experimental category by the CAA, later the FAA, had given flyers an alternative to often-restrictive certification procedures. Flight test requirements were far less complex. The experimental category, incidentally, included racing, exhibition, and pure research and development aircraft, as well as aerobatic and sport airplanes.

A new or emerging factor that expanded the potential of the kitplane segment was increasing business use of homebuilt aircraft. Design emphasis expanded from small single-seat aerobatic or sport models toward more capacious aircraft suitable for business use. Aerodynamics became more advanced. Once again, entrepreneurship in general aviation was enhanced, as expanded roles for kitplanes, including business use, attracted new competitors. Market entry costs for a new firm remained far less than that for entry into factory production of complete aircraft.

Prominent among those attracted to the field by growth and potential for further growth was James Bede. A highly innovative designer of light aircraft, Bede began his company in Springfield, Ohio. The two-seat Bede BD-1, which first flew on July 11, 1963, was developed into the successful American Aviation AA-1 Yankee production aircraft previously described. Bede then developed a series of innovative designs under his BD designation prefix for home construction. Over 2,000 sets of plans for the BD-4 two-seat cabin model were sold. He followed with a developed BD-4 and a four-seat BD-6, and still later with the BD-8 single-seat light aerobatic model.

But the radical BD-5 Micro engulfed the company in controversy. A very small, swept-wing, mid-engined pusher design, the BD-5 was claimed to be capable of 200 mph, even though powered only by a 26 hp engine. It was also claimed to require only 300 hours of assembly time and was promoted as the mass-market small aircraft. The airplane was sold to constructors in a series of sequential subkits, but serious delays in shipments of the engine subassemblies developed, and there were also severe cooling problems with the pusher-engine installation. (Pusher designs did not enjoy the natural engine cooling effect by the propwash of tractor installations.) As a result, the Bede company experienced unstable operations throughout the 1970s, and its reputation suffered. It finally declared bankruptcy in 1979, leaving many builders without the necessary subkits to complete their aircraft.[13] While perhaps unique, the Micro episode served to illustrate the numerous pitfalls of the homebuilt or kitplane business.

The homebuilt dichotomy remained that sets of plans were sold for "scratch" construction and kits of components sold for amateur assembly. Some firms offered both, and some kit producers also offered ultralight/microlight models. There were, in addition, offerings of kit-built autogiros and helicopters, and of light pusher-engined amphibians. Regardless, all kits or plans were marketed on the claim that they were relatively easy to build by those of average skills and that man-hours required were manageable, but the low ratio of actual registered aircraft to plan/kit sales belied the claims. Aviation trade periodicals were replete with advertisements of partially completed kit models for sale by their owners.

The EAA had been founded essentially for experimental designers and builders. But as it grew, the EAA was somewhat surprised to learn that it had effectively inherited a complete segment of the general aviation industry, that of smaller single-engined sport and aerobatic aircraft. That status, however, was largely due to withdrawal from that category by the factory producers, a reaction to product liability costs and overall economics.

A major incentive to growth of the homebuilt segment was the 51 percent rule of the FAA, extending from the CAA. The rule stated that if a homebuilder completed 51 percent of an aircraft, that builder also qualified for a repairman certificate. The practical benefit of the rule was the recognition by the FAA that anyone building 51 percent or more of an airplane also was qualified to service the aircraft and thus could perform 100-hour inspections and other maintenance, realizing major cost savings.

The 51 percent rule also pointed out the trade-off inherent in kit construction, that is, the more complete the kit, with preassembled components, predrilled holes, and precut parts, the less is required of the homebuilder, but at a much higher cost. A rule of thumb developed that for a given amount of cost for work performed by the kit supplier, the cost to the homebuilder is double that amount. Conversely, the less done by the kit supplier, the more that is required by the homebuilder, subsequently with greater potential for error but at much lower cost. Decisions thus were required over such matters as predrilled holes and preassembled components by the kit supplier. The Christen Eagle, a popular small aerobatic model, was threatened with violation of the 51 percent rule, and designer Frank Christensen made changes to increase the amount of home construction required, thus winning approval from FAA inspectors.[14]

In practical terms, the 51 percent rule meant that the builder must complete between 51 percent and 100 percent of the aircraft. Thus a supplier simply supplying plans and a few components that required the amateur builder to build 90 percent would still fall within the 51 percent rule.

A further advantage of the 51 percent rule was that it circumvented product liability for the manufacturer or supplier. If more than 51 percent of an airplane was built by an amateur, the manufacturer of the kit could not be held liable for any design or construction flaws. The rule further protected the kit producers in that factory producers were effectively prevented from offering nearly complete aircraft for which purchasers could finish details for cost savings. Since the factory built more than 51 percent, product liability still applied. Further, such aircraft still were required to be serviced by a certified mechanic.

The contemporary homebuilt segment dates from the mid-1970s. As before, the purely sport or recreational flying segment simply offered little market potential for factory-built aircraft, but rapidly escalating prices of all factory-built aircraft increased market potential for kit aircraft. The additional growth potential of kit-produced business aircraft led to new and advanced kit designs for that purpose. Some featured aerodynamic advances over factory aircraft. Single-engined design by established producers had not progressed significantly since the 1960s, and production declines during the product liability crisis did not aid aeronautical advances. Further, such popular designs as the high-winged Cessna models dated from the 1940s, leaving a potential design gap for kit designers to exploit. A structural advance was the increased use of composites in kitplanes, but composites also possessed drawbacks for small aircraft over traditional materials such as wood or aluminum. Composites had no "give" and tended to shatter on impact, which created a new safety concern.

The homebuilt/kitplane industry segment finally attained major importance in general aviation in the 1980s, appearing to join the mainstream. The major market opportunity for competitors in the segment lay with the decline or virtual disappearance of factory-produced light aircraft, especially small trainers. Whatever market existed for small single-seat sport or aerobatic aircraft was filled, as before, by the kitplane suppliers, as those models had never been a significant factor in factory-built aircraft. The homebuilt sector thus helped preserve the sport/recreational segment and also served as a partial offset to the decline in factory production. Business aircraft use, also price-sensitive, enabled homebuilt business aircraft to grow, and even homebuilt agricultural aircraft appeared. In addition, export potential for kits increased, especially to Europe, where general aviation operations, including fuel, remained substantially more expensive than in North America.

Choice of designs for home construction was widespread and the price range was also broad, but one problem remaining for potential amateur builders, and for the growth of the segment, was that the business was unstable; kit suppliers went out of business regularly. And as a growth industry, most active firms were of recent origin. The average age of a kitplane company was only six years, and designs were traded among firms frequently. Yet maintenance of stable demand or strong market growth required a stable industry. The segment also possessed its international aspects, as European and other foreign designs were licensed to U.S. firms. In addition, the small Rotax engine from Austria, a light, two-stroke model in the 50 hp class, was a popular powerplant for the smaller homebuilts.

The 1995–96 edition of *Jane's All the World's Aircraft* listed no fewer than 36 suppliers of plans or kits (or both) in the United States, although many more existed. The following survey briefly describes the more important firms. Among the most important competitors was Stoddard-Hamilton Aircraft, Inc., of Arlington, Washington, founded in 1979. Stoddard-Hamilton offered the Glasair line, originally designed by Tom Hamilton, with a structure of molded composites and advanced aerodynamics. The original Glasair flew in 1979, and the Glasair II-S and III succeeded the earlier model, offering retractable gear and higher performance. Representing the high end of the homebuilt field in performance, the new Glasair Super II offered 250 mph speed and was available in retractable gear, fixed tricycle, and taildragger gear configurations. Broadening its coverage, the company introduced in 1996 the GlaStar, an entry-level small model, with a kit priced at $19,000.

Europa Aviation, Inc., of Lakeland, Florida, offered the Europa line. The Europa, a simple low-wing taildragger-gear model powered by an 80 hp Rotax, was designed and developed in the United Kingdom. The U.S. subsidiary marketed the models domestically.

Neico Aviation of Santa Paula, California, offered the Lancair line, another design with advanced aerodynamics. The current models 200 and 235 featured a Kevlar composite structure. The first Lancair flew in June 1984. The Lancair IV was claimed to be the fastest kitplane, and its promise led the company toward full factory production. Other models in the line included the 360, ES, and Super ES, which was developed as the LC40, a fixed-gear four-seat design. Volmer Aircraft of Glendale, California, offered the VJ-22 Sportsman amphibian. Volmer Jensen, a noted sailplane designer, broadened his activities into amphibians.

Possibly the most promising firm was Cirrus Design Corp. of Baraboo, Wisconsin. Cirrus was founded originally for kit production but relocated to Duluth, Minnesota, with plans to move into full factory production of its designs. President Alan Klopmeier became a leading figure in the field. The first Cirrus Design model to receive FAA certification was the ST-50, developed in collaboration with Israviation of Israel. The ST-50, a five-seat high-performance pusher turboprop for business use, went into production in Israel. The domestic Cirrus Design VK30 advanced pusher-engined 4–5 seat model first flew on February 11, 1988. The initial kit price was $42,500, including engine, propeller, and avionics, but the VK30 program suffered a setback when a test pilot was killed in a crash on March 22, 1996.

An entirely new Cirrus Design model, the SR-20, was developed beginning in 1990. A modern, economical, low-wing four-seat all-composite model for factory construction, the SR-20 projected an initial base price of $130,000. Major components are built at a new factory in Grand Forks, North Dakota. The state had enacted laws to head off aircraft product liability suits, which attracted Cirrus.

The Scaled Composites, Ltd., firm, founded by Burt Rutan in 1982, continued to be a factor in the homebuilt field, although its activities were broader. Scaled Composites was a separate venture from the original Rutan Aircraft Company formed in the 1960s. The Rutan-designed VariEze was begun in 1974 and first flew on May 21, 1975. The larger, more powerful LongEZ first flew on June 12, 1979. It became a popular homebuilt model for cutting-edge pilots. Both were radical composite designs with canard configurations. Rutan had worked for Bede earlier, and much Bede influence was evident in his emphasis on composites and advanced aerodynamics. Rutan continued as a leader in advanced concepts, especially with the VariEze, a canard with aft-mounted engine which avoided the cooling problems experienced by the BD-5. The design emphasized stability as well as high performance. Rutan operated Scaled Composites independently after reacquiring it from Beech in November 1988.

Another growing firm was Aviat Aircraft, Inc., of Afton, Wyoming. The company originated in California as Christen Industries. Frank Christensen designed the Christen Eagle, a small, fully aerobatic biplane, which competed with the designs of Ray Stits. Christensen earlier had attempted to purchase rights from Stits, who declined to sell. The Eagle kit was originally in the $25,000 class. Curtis Pitts had operated at Afton, and Christen instead acquired Pitts Aerobatics rights in November 1983, but later ceased marketing plans or kits. The Christen Eagle II, a high-performance model with a 200 hp Lycoming and intended for kit construction, made its first flight in February 1977. The Eagle II led to the formation of Christen Industries in 1984. Curtis Pitts remained active in experimental aircraft.

Christen was acquired in April 1991 by Aviat, a subsidiary of a British company, and took the name Aviat Aircraft, Inc. Malcolm White became the new owner and Frank Christensen retired. The Aviat A-1 Husky, a factory-built two-seat light utility aircraft, designed by Christensen in 1987, enjoyed steadily increasing sales. In fact, the Husky bid to replace the veteran Super Cub in utility roles. Aviat also offered factory-built Pitts models S-1T, S-2B, and S-2S to order. The Eagle II remained on the market, but kit production was a minority of total Aviat business. While remaining popular, the current Eagle II kit rose in price to $55,000, less engine and propeller. Aviat operated in the old Call facilities, which it expanded over time. New investor Stuart Horn purchased Aviat in December 1995 and further expanded operations. Aviat recently acquired the type certificate to the old Globe Swift from Piper, with plans to restart production in the future.[15]

SkyStar of Nampa, Idaho, took over the popular KitFox design, which first flew on May 7, 1984, from Denney Aerocraft Co., its original developer. *Fortune* reported in 1983 that the KitFox, a simple side-by-side, high-winged STOL model, was the best-selling kit model, with 2,000 sold from 1984 to 1993. Denney more recently was reported to have purchased the Pulsar line from Aero Designs of San Antonio, Texas. The basic Pulsar, a low-wing two-seat modern design with Rotax power, carried a kit base price of $21,500, rising to $27,500 for the more powerful Pulsar XP.

Sequoia Aircraft Corp. of Richmond, Virginia, was founded in 1975. Sequoia developed its original Models 300 and 302 for kit construction, but a more significant move was the acquisition of rights to the Italian Falco F.8L. The Falco was a progression of an original design from the engineer Stelio Frati. It first flew in 1955. A modern two-seat retractable-gear low-wing model, Sequoia adapted the Falco for home construction with a basic kit cost of $60,000.

Aero Commander 560 production line, Bethany, Oklahoma, 1950s. (*Source: National Air and Space Museum, Smithsonian Institution.*)

Clyde Cessna and Dwane Wallace standing beside a Cessna 180, 1953. (*Source: National Air and Space Museum, Smithsonian Institution.*)

W. T. Piper, Sr., circa 1953. (*Source: National Air and Space Museum, Smithsonian Institution.*)

Piper Apache production line at the new assembly building, Lock Haven, Pennsylvania, 1956. (*Source: National Air and Space Museum, Smithsonian Institution.*)

Ceremony for the completion of the 30,000th Beech aircraft, circa 1969. Left to right: Frank Hedrick, Olive Ann Beech, and John H. Batten, President, Twin Disc, Inc. (*Source: National Air and Space Museum, Smithsonian Institution.*)

Custer Channel Wing prototype, 1964. W. R. Custer, at right, with his wife and other family members. (*Source: National Air and Space Museum, Smithsonian Institution.*)

Two Bell Model 47 helicopters engaged in logging operations in Canada during the 1950s. (*Source: Bell Helicopter Textron.*)

The classic Pitts S-2B aerobatic biplane, now built to order by Aviat Aircraft. (*Source: Aviat Aircraft, Inc.*)

The Sikorsky S-76, the largest general aviation helicopter, in the Emergency Medical Service configuration. (*Source: Sikorsky Aircraft.*)

A new Bell Model 412 accompanied by a Bell Model 206 flying over Dallas. (*Source: Bell Helicopter Textron.*)

The Aviat Husky A-1, a successful light utility aircraft. (*Source: Aviat Aircraft, Inc.*)

The classic Pitts S-2B aerobatic biplane, now built to order by Aviat Aircraft. (*Source: Aviat Aircraft, Inc.*)

The Sikorsky S-76, the largest general aviation helicopter, in the Emergency Medical Service configuration. (*Source: Sikorsky Aircraft.*)

Ceremony for the completion of the 30,000th Beech aircraft, circa 1969. Left to right: Frank Hedrick, Olive Ann Beech, and John H. Batten, President, Twin Disc, Inc. (*Source: National Air and Space Museum, Smithsonian Institution.*)

Custer Channel Wing prototype, 1964. W. R. Custer, at right, with his wife and other family members. (*Source: National Air and Space Museum, Smithsonian Institution.*)

Two Bell Model 47 helicopters engaged in logging operations in Canada during the 1950s. (*Source: Bell Helicopter Textron.*)

William P. Lear, Sr., aviation entrepreneur and originator of the Learjet series. (*Source: Learjet, Inc.*)

The prototype Learjet Model 23 taking off on its first flight, October 7, 1963. (*Source: Learjet, Inc.*)

B. D. Maule, founder-president, Maule Air, Inc., 1991. (*Source: Maule Air, Inc.*)

The Ayres Turbo Thrush S2R-T34, a powerful agricultural aircraft. (*Source: Ayres Corp.*)

The Ayres Turbo Thrush S2R-G10, with the P&W Canada PT6A engine. (*Source: Ayres Corp.*)

The sixth Learjet Model 45, on its first test flight over Wichita, November 6, 1997. (*Source: Learjet, Inc.*)

President Clinton signs General Aviation Revitalization Act, August 17, 1994. (*Source: White House photo office.*)

A successor Bede firm located in St. Louis, Missouri. James Bede remained active in advanced developments with the advanced BD-12 and BD-14 pusher-engined models which continue the basic concept of the ill-fated BD-5. Both were powered by an 80 hp Rotax. Current BD-12 kit prices range from $21,900 to $40,000, less engine, propeller, and instrumentation. A further marketing innovation was that Bede announced plans to market kits through upscale automobile and boat dealerships.

Bede also designed a BD-10 civil two-seat supersonic personal craft, production rights for which have been sold to Peregrine Flight International of Nevada. Peregrine planned full factory production but also planned to offer the BD-10 in kit form. Completed cost was expected to be approximately $1.4 million.

Smaller competitors included Rans, Inc., of Hays, Kansas, and Zenith, of Mexico, Missouri. Rans developed the Coyote series of small high-wing models and offered ultralight models as well. The new S-7 Courier, with a kit price of $15,300, appeared to be a contender. Zenith began with designs acquired from Canada.

Among older names, Taylor Aerocar, Inc., after deciding that regulations and economics had effectively blocked flying automobile production, attempted an entry in kit business for its roadable airplane. David B. Thurston designed and offered the TA-16 Trojan 4-seat pusher amphibian as a kit, but sales have been limited.

A more recent factor in the sector, and gaining in significance, was the increasing crossover between traditional factory builders and kitplane or plans suppliers. Aviat, Cirrus Design, and Bede were prominent examples. But there was no evidence of the reverse, that factory builders were interested in offering kits.

Numbers of amateur-built aircraft on the civil registry grew fairly steadily. Registrations totaled 2,865 in 1971, growing to 7,496 by the end of 1981. By 1985 the number exceeded 10,000 and grew to 19,104 registered by July 1, 1997.[16] In addition, there were several hundred racing aircraft on the registry. Attendance at annual EAA Fly-Ins grew to more than 800,000 people. While the actual number of homebuilt aircraft flying remained difficult to track, it was determined that by 1990 the number of kits supplied to homebuilders exceeded shipments of factory-built small aircraft by at least two to one.[17] A key role of the EAA was assisting aspiring homebuilders in their task.

It remained difficult to compile annual values of aircraft built from kits and plans due to the varying value added by homebuilders and to high noncompletion rates. One recent estimate was that the completion rate from kits was 63 percent, while that from plans was less than 5 percent. But there was no question that the impact of the experimental and homebuilt segment on general aviation was rising. It was generally recognized that the homebuilt segment had effectively replaced the traditional factory-built trainer segment.

Helicopter Survey

The civil helicopter market suffered a serious decline from 1980, even as helicopter capabilities advanced and functions expanded, and most competitors struggled. The helicopter had acquired increasing importance in traditional bush flying and in logging; Canada, in fact, was the second-largest civil helicopter market in the world. Reasons for the market decline were many, but mainly included the slump in offshore oil activity, previously a strong growth area; sharply rising product liability costs; expense and availability of operator's insurance; a large fleet of good used helicopters; noise and safety concerns; and continuing high operating costs.[18] A further limitation in demand for new helicopters appearing from the late 1980s was the increased release of smaller surplus military helicopters to the civil market. Military downsizing meant, among other things, an excess helicopter inventory, and many were similar to civil models. Police forces in particular were beneficiaries. Yet development continued to be active, and competitors determined to enter or remain in the market, as the following survey recounts.

Brantly-Hynes, in another name change, became Hynes Helicopter, Inc., in 1984, but with declining business the firm closed and all assets were put up for sale in 1987. Then new interests purchased all rights for the B-2B and Model 305 from Hynes Aviation Industries in March 1989. Brantly Helicopter Industries was incorporated on May 8, 1989, in Texas, with plans to restart production, but the venture eventually faded.

In January 1980, F. Lee Bailey sold Enstrom to Bravo Investments, B.V., of the Netherlands, which continued production. Then in September 1984 Bravo resold Enstrom to a new American investor group which further developed the turbine-powered five-seat Model 480. A disappointment was the loss by the TH-28 military version of the Army New Training Helicopter (NTH) order, but civil production continued.

Robinson continued to lead the light piston-powered helicopter segment. The company delivered its 1000th helicopter on March 30, 1989. In the recession year of 1990, a conspicuous success was the delivery of 402 R22 models, and total production reached 2,260 on November 30, 1992. Robinson announced the four-seat R44 in 1989, still relatively inexpensive, which won 50 orders by 1992. Robinson emphasized the export market for the R44 to avoid the product liability situation in the domestic market.

Hiller Aviation finally acquired rights to the Fairchild FH-1100 in April 1980. Hiller could not finance a resumption of series production, however, and suffered bankruptcy on January 23, 1984, leading to its sale to the Rogerson Aircraft Corporation in April 1984. Hiller Helicopters then operated as a wholly owned subsidiary of the renamed Rogerson-Hiller Corporation. The firm relocated to Port Angeles, Washington, where it won some subcontracting orders from Boeing, but helicopter production remained suspended.

The sailplane manufacturer Schweizer, having purchased all Ag-Cat rights from Grumman in January 1981 and continuing production of an advanced version, acquired rights to the Hughes 300 civil model on July 13, 1983. Schweizer thus became more heavily committed to powered aircraft. Hughes sold its civil line after declining demand, and Schweizer produced the Model 300 primarily for the training market. The developed business Model 330, with a base price of $517,000, also entered production. The Model 330 was also entered into the Army NTH competition but was eliminated.

Bell Helicopter Textron (BHT) was incorporated as a wholly owned subsidiary of Textron, Inc., on January 3, 1982. BHT then established a wholly owned Bell Helicopter Canada (BHC) subsidiary at Montreal in October 1983 to produce for the civil market and enable the parent to concentrate on military production in Texas. The investment responded to incentives provided by the Canadian government.

Production of the JetRanger line began to be transferred to the new location during 1985. The larger Models 212 and 412 were transferred to Canada in mid-1988 and 1989, respectively, and the Model 206LT Light Twin was added to the 206 series in 1992. The Model 412 also was produced under license by Agusta in Italy. The 412 was the intended successor to the 212, but the 212 remained in production as demand appeared steady. But initially strong demand for the Model 222 intermediate twin, still produced in Texas, faded in the general market decline, and production ended in 1988 with 182 built. The improved successor Model 230, first flying on August 12, 1991, also sold slowly in the sluggish market, and production ended in 1995 at 38.

Helicopter development was at both Fort Worth and Montreal, but approximately half of all Canadian components were supplied from Fort Worth. All BHC helicopters, with the exception of those for the Canadian market, were sold to the Texas parent for resale. A major order came in 1991 for 66 JetRangers and 88 LongRangers from Petroleum Helicopters (PHI). The largest fleet operator in the world, PHI needed the craft for oil service operations, primarily in the Gulf of Mexico.[19] Another important step for BHC was the Army NTH competition, won by the TH-67 Creek, which was derived from the JetRanger III. An initial order for 102 was placed in March 1993. The TH-67 was built in Canada as well.

In March 1982, Sikorsky began deliveries of the improved S-76 Mark II. The largest general aviation helicopter, the S-76's capacity and performance was increasingly favored over smaller models for servicing oil drilling platforms despite its $6–$7 million price.

McDonnell Douglas acquired Hughes Helicopters of Mesa, Arizona, from the Hughes estate in January 1984. The operation was renamed McDonnell Douglas Helicopter Company (MDHC) on August 27, 1985 and headquartered at Mesa. While primarily involved with military attack helicopters, MDHC became increasingly oriented toward the civil market. The civil Model 500, developed from the LOH design, won some orders. Then the advanced NOTAR (no tail rotor) development, with lateral control by deflected air flows, resulted in the MD 520N (NOTAR) of 1989, based on the Model 500 but extensively redesigned. It was funded by a defense contract. The MD 520N held promise for civil markets as well as military applications, with the first civil delivery occurring in November 1991.

The civil helicopter segment held problems in common with other segments, but a basic strength was that in 1984 the civil helicopter fleet numbered some 7,000, including those owned by public service or governmental organizations. In addition, U.S. producers accounted for some 60 percent of the world market.

Product Liability

Product liability lawsuits, in which manufacturers were held financially liable for crashes and resulting injuries and fatalities due to findings of design or production flaws, began to devastate general aviation aircraft production by the early 1980s. Although reflecting in part the overall litigiousness of American society, the trend had a greater impact on general aviation than on other, larger industries. Product liability became defined so broadly that even an engine failure could be ruled a design or manufacturing defect, although it had always been recognized in aviation that any man-made engine could fail, however rarely. For that reason, no single-engined commercial airliners were allowed to operate. Increasing court awards began to inhibit production of single-engined models, always the heart of general aviation and numerically the most prominent. This in turn precipitated a downward spiral in which enrollment in flight schools, always expensive, declined further, leading to decline in demand for new trainers. Light trainer production in the United States ended entirely.

The crisis developed slowly. By 1971, it was realized that liability costs to the manufacturers had increased fifteenfold over the decade. Previously amounts were nominal by comparison. Bill Piper, Jr. first raised the product liability concern with GAMA in that year. Product liability suits appeared with increasing frequency during the 1970s but appeared manageable at the time. Beech, for example, after defending itself against several suits, won a major class action suit in federal court in January 1973, in which all pending product liability litigation against the firm was dismissed.[20]

By the early 1980s, however, product liability had mushroomed into a major crisis. Insurers began to base premiums not on the value of the airplane but on the potential magnitude of a court liability award, and liability insurance became so expensive that it threatened the financial viability of general aircraft manufacturers. As more lawsuits occurred, the broad societal choice, as represented by juries, was toward the manufacturer as the provider of relief. Attorneys for the manufacturers increasingly advised out-of-court settlements when possible as the more expeditious solution, but that in turn damaged the firms' reputations for quality.[21]

The average manufacturer's liability insurance premium per single-engined plane rose from about $2,000 in 1972 to the $60,000–$100,000 range by 1988, often doubling the cost of a single aircraft.[22] Actual cash outlays by the industry for court judgments, settlements, and defense costs soared tenfold, from $24 million in 1977 to $210 million in 1985.[23] Moreover, there was no time limit; the "liability tail" extended to all aircraft ever manufactured by a firm, which in some cases covered more than 50 years. A new owner of a general aviation firm thus faced potential liability for aircraft built by the company decades before, and resulting insurance costs were reflected in the price of new aircraft. Furthermore, the fact of FAA certification was no defense against claims of design flaws, which ironically served to inhibit product improvements in that they might be interpreted as an admission by the manufacturer that it was correcting a design flaw.[24] Despite the deepening problems, manufacturers won most product liability suits, but the litigation expense and massive awards involved in those lost still proved to be nearly devastating.

Manufacturers in general pressed for imposition of a Statute of Repose for product liability, advocating a time limit of 10 to 12 years as reasonable. But legislative relief encountered strong opposition, particularly from the Association of Trial Lawyers of America (ATLA). An early bill stalled in the U.S. Senate in 1985. It appeared that the general aviation industry simply lacked the political influence necessary to obtain legislative relief.

John W. R. Taylor, editor of the authoritative *Jane's* reference publication, in his preface to the 1989–90 edition, characterized the general aviation problem as "crippling and ludicrous liability litigation waged by unscrupulous lawyers." He went on to praise new Piper owner M. Stuart Millar's attempt to fight such suits in the courts with staff attorneys rather than relying on insurance.[25]

In addition to product liability, unfavorable changes in federal tax laws, including lower allowable depreciation for business aircraft and elimination of the investment tax credit (ITC) in 1986, previously applicable in new business aircraft purchases, further diminished general aviation sales. Costs of replacement models, driven by liability insurance, were sharply higher than earlier models, leading to declining demand for new aircraft. Airline deregulation from 1978 also resulted in a long-term deleterious effect on the general aviation market, in that the expansion of third-level airlines opening or resuming service to smaller and more remote communities lessened the need for many business users to own and operate their own aircraft. High interest rates from 1979 were another factor limiting demand.

The general aviation industry bore some responsibility for its decline, however, in that technology and performance of its current models remained similar to those of the late 1950s and early 1960s. In fact, many aircraft in production in the 1990s were developments of 1960 and earlier designs. While there was continuous product refinement, no technological revolution in general aviation aircraft had occurred, and critics charged that the industry followed a cookie-cutter approach to design and production. Yet design conservatism may have been partially dictated by certification difficulties often attending a radically new design. Certification had delayed the Beech Starship, and certification problems also weighed against the innovative Lear Fan 2100, as had been the case earlier with the Custer channel-wing design. Another complication was that general aviation aircraft were so durable and reliable that with proper maintenance they could last indefinitely. Operators simply felt no need to trade in aircraft every three to five years as automobiles might be replaced. There was no planned obsolescence in aircraft.

The industry also began to realize that somewhat reminiscent of the market collapse of 1947, the decline was due in part to overproduction in the 1970s. Current production had been in excess of underlying demand. One stimulus was that as prices increased rapidly with inflation, speculative buying became a major factor, as buyers in many instances resold their aircraft later at a profit.[26] The market simply became saturated. A decline in demand, particularly in the single-engined segment, inevitably would have occurred, even without the product liability crisis.[27] The general aviation sector also appeared to have insufficient competition; it was definitely oligopolistic, with a four-firm concentration ratio of above 85 percent of total business.[28] A possible explanation was that smaller firms simply could not acquire the financial strength and product line diversity needed to compete in the product liability environment against the larger, financially stronger firms.

The general aviation industry, despite myriad problems, still had not collapsed or been completely destroyed. One view was that it had instead undergone, albeit at great distress, a fundamental structural and technological transformation.[29] As one manifestation, after production of small single-engined trainers had been halted, the kitplane segment took off in direct consequence, effectively replacing that factory-built segment and enabling thousands to continue sport flying. Kit aircraft for amateur construction made flying and ownership more affordable in an unfavorable economic environment. In addition, improved kitplane technology enabled the segment to move more closely to the mainstream of general aviation.[30] Growing political influence of the EAA and increasing FAA support for homebuilts also helped.

Table 7-3
Civil Aircraft Shipments, 1980–1995
(Figures are consistent with totals in Table 7-2)

Year	Total	Helicopters	General Aviation
		Quantity	
1980	13,247	1,366	11,881
1981	10,529	1,072	9,457
1982	4,853	587	4,266
1983	3,094	403	2,691*
1984	2,814	376	2,438
1985	2,413	384	2,029
1986	1,825	330	1,495
1987	1,443	358	1,085
1988	1,526	383	1,143
1989	2,050	515	1,535
1990	1,747	603	1,144
1991	1,592	571	1,021
1992	1,223	324	899
1993	1,222	258	964
1994	1,236	308	928
1995	1,369	292	1,077

Year	Total	Helicopters	General Aviation
		Value ($ millions)	
1980	$3,163	$656	$2,507
1981	3,517	597	2,920
1982	2,364	365	1,999
1983	1,773	303	1,470*
1984	2,028	330	1,698
1985	1,937	506	1,431
1986	1,550	288	1,262
1987	1,641	277	1,364
1988	2,252	334	1,918
1989	2,055	251	1,804
1990	2,262	254	2,008
1991	2,179	211	1,968
1992	1,978	142	1,836
1993	2,257	113	2,144
1994	2,542	185	2,357
1995	3,036	194	2,842

*Includes three off-the-shelf Gulfstream IIIs delivered to the U.S. Air Force as C-20 VIP transports.

SOURCE: *Aerospace Facts and Figures, 1996/1997,* p. 32.

The general aviation industry, as did the much larger military aerospace sector, regarded itself as a national resource, worthy of societal and political support, especially in a period of crisis. But real stability and prosperity could come only from market expansion, unlikely given the unfavorable legal and economic environments and the pricing structure of general aviation aircraft. Ironically, market problems appeared as private flying became steadily safer, with fatalities extremely low in relation to hours flown.

Dramatically illustrating the market decline (see Table 7-3), GAMA reported that shipments by its 10 manufacturing members totaled only 2,691 aircraft in 1983, the lowest total since 1951, although the value of those aircraft was almost $1.5 billion.[31] The decline in single-engined aircraft production from more than 14,000 in 1978 to only 613 in 1987 further underscored the crisis.[32] Employment in general aviation manufacturing was down to 21,000 by 1985. Cessna commercial aircraft sales declined to 1,217 in 1983 and to only 978 in 1984, down from 6,393 in 1980, and the company sustained a financial loss in 1983.

By 1990 general aviation production was only a shadow of its boom days of the 1960s and 1970s. GAMA members delivered only 1,021 aircraft in 1991, in an environment of overall recession.[33] The civil helicopter market was also depressed. The decline led manufacturers to market and distribute their products directly, weakening their well-established dealer network.

One enduring strength of the U.S. general aviation industry was that it remained effectively the world general aviation industry, since no other country possessed a comprehensive production and market capability in the field. But exports could not offset the domestic decline. Export demand, while always a significant factor, declined as sharply as domestic demand due to economic problems in other markets comparable to those in the United States. In addition, the U.S. dollar, generally strong in relation to other currencies during the 1980s, did not help exports to remain competitive.[34]

# # # # #

In the midst of a crisis in the general aviation industry, the last remaining major flight record, an unrefueled nonstop around-the-world flight, was achieved during December 14–23, 1986. The Voyager design of Burt Rutan, essentially a general aviation craft powered by push-pull Lycoming engines and piloted by Dick Rutan and Jeana Yeager, covered more than 25,000 miles. Rutan lacked major financial support, building and testing the craft with his own resources. The feat recaptured the pioneer spirit of record-setting flights by adventurous aviators, particularly significant during the product liability crisis.

8

The Path to Recovery

By the end of the 1980s production of general aviation aircraft had declined to the point that the industry was practically moribund. The year 1985 saw not only twin-engined Aero Commander production ended and Learjet production suspended, but also production shutdowns of numerous smaller models. Piper closed its Lock Haven plant in August 1984 after completing the last of more than 76,000 aircraft at that location.[1] The Lakeland factory closed in October 1985. Beech closed its Liberal, Kansas, plant in August 1985, and soon thereafter closed its Selma and Boulder facilities. Largely out of light business aircraft by that time, Beech consolidated production at a single factory, as had Cessna earlier. The remaining smaller firms essentially produced to order. Business jet sales experienced a modest recovery at the end of the decade, but it was not to endure.

Yet in the face of seemingly overwhelming negative prospects about the general aviation industry and its future, there was no question of its economic importance. Estimates for 1990, for example, were that general aviation contributed $40 billion to the Gross National Product, accounted for 540,000 jobs, and served 120 million people in the United States in some manner.[2]

Company Developments

Russ Meyer remained CEO of Cessna after the General Dynamics acquisition. He moved immediately to suspend production of all piston-engined aircraft. The Conquest I and II were the last of Cessna's once-extensive twin-engined line to remain in production, but all aircraft other than the Citation series and the Caravan were suspended indefinitely in May 1986. By that time, however, Cessna had built an extraordinary record, producing 35,772 of the 172/Skyhawk/175 Skylark series, 31,340 of the 150/152 series, and 19,812 of the 180/182/Skylane series. Except for the Models 150/152, the figures included those produced by Reims Aviation. In a further contractionary move, Cessna sold in February 1989 its 49 percent interest in Reims, which it had held for almost three decades.[3] Reims had produced more than 6,000 aircraft of Cessna design.

Cessna was more vulnerable than Piper to product liability awards because of the resources of its large corporate parent, which hastened the suspension. Employment fell from more than 14,000 to 3,600 by 1985. Russ Meyer, probably the leading spokesperson for the general aviation sector, stated that his company would not resume production of single-engined models until product liability laws were reformed.[4]

Cessna continued business jet leadership, with the Citation series providing continuing profits for the company before and after suspension of other models. The Caravan utility aircraft also continued strong sales, with over 400 produced by 1990. Federal Express remained the largest customer. The lengthened Model 208B Grand Caravan, with a capacity of 14 passengers and priced in the $1 million range, entered production in September 1990 and also found customers.

Mooney continued with its single-engined four-seat line, adding the advanced MSE in 1989 and the turbocharged M-20R Ovation in 1994. Overall production averaged some 100 aircraft per year, and in 1995 Mooney passed the production milestone of 10,000 aircraft in its 47-year history. Mooney entered into a multinational joint venture with the Socata subsidiary of the French national firm Aerospatiale on June 12, 1987, to develop the TBM 700, a large single-turboprop pressurized business aircraft. The venture enjoyed 30 percent French government financing, but Mooney withdrew in May 1991. Socata continued the program with the Finnish firm Valmet, and the TBM 700 won limited orders in the U.S. market.

Only Beech approached a full line of business aircraft, though it no longer produced small trainers. Since the majority of product liability suits involved single-engined and training models, a manufacturer needed the cushion of larger business aircraft to sustain production in single-engined aircraft. Beech had that cushion and thus was able to offer a broad product line. Also, manufacturers were induced to concentrate on large executive aircraft because corporate operators possessed greater maintenance resources, lessening the likelihood of accidents and consequent product liability suits. Liability insurance also was proportionately less costly for more expensive models.

Beech in 1991 offered the continually upgraded King Air C90 and Super King Air series, the Beechjet 400A, the Starship 2000, and the Model 1900D. After closing the Selma plant in September 1986, Beech transferred Model 1900 production to Wichita. Beech delivered 255 Model 1900Cs between 1984 and 1991, and the type was largely responsible for Beech's improved earnings in 1988 and 1989, even in the face of the crisis in general aviation. The 1900C was succeeded by the improved Model 1900D, with a stand-up cabin, which first flew on March 1, 1990. The Model 1900D, while priced in the $5 million range, continued to sell in its segment. The piston-powered straight-tailed Bonanza and Baron also continued but constituted only a small percentage of Beech's total sales. May 1996 marked the milestone of more than 3,000 straight-tailed Bonanzas produced since their 1968 introduction. The Bonanza series marked its 50th anniversary in 1997, a world record for continuous production.

The Super King Air became the most successful and best-selling twin turboprop ever produced, with more than 2,000 of the series delivered from 1973. June 1996 marked the milestone of 5,000 King Air/Super King Air deliveries, including all military and export models. The export market remained strong, and almost 10 percent of sales were to the U.S. military. Beech achieved another major success when the Beechjet 400T won the Air Force Tanker Transport Training System (TTTS) order on February 21, 1990, over vigorous competition from the Learjet and the Citation S/II. Designated the T-1A Jayhawk, 180 were ordered for service from 1992.[5] Beech sales were second only to missiles and electronics among Raytheon business lines.

Max Bleck, president of Beech Aircraft, became president of parent company Raytheon on March 1, 1991. Arthur E. Wegner, formerly with United Technologies and a former president of Pratt & Whitney, became Beech chairman and CEO in July 1993. Olive Ann Beech, the most prominent woman executive in the history of the industry, died on July 6, 1993, at the age of 89. Frank Hedrick had died in

1984. Beech completed production of its 50,000th aircraft early in 1993, particularly noteworthy given its historical focus on the higher end of the general aviation market.

Piper continued to contract operations with the market decline, consolidating remaining production at Vero Beach, where a new factory was completed in October 1986. The firm's brightest hope was its PA-46 Malibu turbocharged, pressurized, six-seat executive model, which gained a strong market share.

The California industrialist and aviation enthusiast M. Stuart Millar purchased Piper on May 12, 1987, at a distress price, from Forstmann Little, which had just purchased Piper from the bankrupt Lear Siegler Corporation. Millar placed the company under his Romeo Charlie investment firm, in effect taking the company private.[6] Millar rationalized the product line, built a strong order backlog, and resolved to fight crushing insurance costs by hiring a staff of lawyers to fight product liability suits in court. He thus hoped to reassure distributors and preclude future litigation. But by purchasing Piper, Millar also became potentially liable for every aircraft built by Piper over 50 years.

To revive the trainer market, he began production in 1988 of the two/four seat PA-28-161 Cadet, derived from the Warrior II. To hold down cost,[7] he proceeded without adequate liability insurance. He also resumed production of the veteran Super Cub in 1988 after WTA had withdrawn as licensee and continued the license production agreements in Argentina, Poland, and Brazil. Major production activity was with the Cheyenne IIIA; the developed Malibu Mirage, which appeared in October 1988; the Seneca; and the Cadet, but as conditions worsened, only the Malibu Mirage remained in series production. Piper engineers also assisted with development of the Chilean ENAER Pillan military trainer, derived from the Cherokee. Millar was determined to expand aggressively, even with a backlog of liability suits.[8]

Another Millar effort was the LoPresti Piper Aircraft Engineering Company, established in December 1987. Ownership was 75 percent by Millar and 25 percent by LeRoy LoPresti, formerly with Mooney. LoPresti also managed the venture. The intent was to produce a new version of the old Globe GC-1A Swift as the SwiftFury two-seat sportplane, but the venture closed in December 1990 due to financial problems. The PiperNorth operation, reestablished in November 1989 at Lock Haven with a Pennsylvania loan guarantee, also closed.

By the fall of 1989 Piper's problems were so severe that suppliers began to cut off credit, seriously disrupting production. The low-priced Cadet, serving as a loss leader, did not help cash flow.[9] The company desperately needed more sales of its Malibu Mirage, priced in the $500,000 range, but the cash crunch forced a two-week shutdown in February 1990. Millar again considered reopening the Lock Haven plant to produce both the Navajo and the SwiftFury, but financial conditions precluded the proposal.

By 1991 it was apparent that Millar's gamble was failing, as continued credit restrictions from suppliers severely limited production and left the company unable to fill orders. New investors were sought, but a prospective sale of Piper to Aerospatiale stalled over concerns about product liability: any new owner faced potential product liability for more than 20,000 Piper aircraft in operation in the United States. A further difficulty was a series of seven Malibu/Mirage fatal accidents during 1989–1991. Eventually, the aircraft's structure was exonerated, but that finding came too late as Piper, crushed by product liability awards and failing to find added financing, filed for Chapter 11 bankruptcy on July 2, 1991. The future of the Piper nameplate after 54 years appeared dim.

A new investor, Stone Douglass, gained control from Stuart Millar by acquiring Millar's Romeo Charlie shares. Operations continued at sharply reduced levels as the search continued for either a buyout or new financing. The company even briefly considered selling rights to Canada and restarting production there in order to escape U.S. product liability. Canadian law was more lenient in the product liability area. The eventual ownership transfer came at the end of 1994, however, to unsecured creditors led by engine supplier Teledyne Continental. Sales were hardly spurred by bankruptcy, but management continued to seek approval from creditors to reorganize and emerge from bankruptcy as a public firm.[10]

On July 17, 1995, new Piper president and CEO Chuck Suma announced a successful restructuring as The New Piper Aircraft Inc., owned by major creditors and with investment bank backing. Piper assets were valued at $95 million. Production of established models, led by the Malibu Mirage, moved ahead. The PA-32 Saratoga and the PA-28 single-engined family also remained in production. The only Piper twin, the PA-34 Seneca IV, was an important factor and was succeeded by a refined Seneca V. Pending market demand, the turbine-powered Cheyenne and Navajo also stood to be reinstated. Piper regained its former position as a market leader.

Fairchild, in San Antonio, continued with low-volume production of its Metro series. Some 1000 had been produced by 1996, and several military variants aided stability. Fairchild began development of a stand-up fuselage version, largely to meet the challenge from its major competitor, the Beech 1900D.

Socata, the Aerospatiale general aviation subsidiary founded in 1966, developed modern small single-engined designs. Product liability concerns had driven most American producers from that segment, opening a window of opportunity for foreign competitors, including Socata. The Socata Trinidad, Tobago, and Tampico made inroads into the U.S. market from the 1980s. In 1995 Socata also acquired rights to the former Grumman American/Gulfstream Cougar light twin.

Even in the troubled environment, the entrepreneurial spirit, which had always characterized the general aviation industry, remained as active as at any time in its history. New ventures appeared, producing either new designs or older models acquired from others, and aspiring to gain a market foothold whatever the current adversity. One such venture in single-engined aircraft was the American General Aircraft Corporation. In June 1989 investors led by James E. Cox, founder and president, purchased type certificates for Grumman American designs from Gulfstream Aerospace, which had ended production of the line in 1980. With an investment by Gulfstream, American General entered production at the Greenville, Mississippi, factory vacated by Boeing in 1990. Priced in the $100,000 range, a refined model of the fixed-gear AA-5B Tiger achieved some initial market success, but low volume forced the firm into a production suspension in 1993 and then into Chapter 11 in February 1994.

New investors acquired rights from Gulfstream to the Commander 112 and Commander 114 of the old Rockwell line from Gulfstream in the summer of 1988. Production of the Commander 114 was revived by the Commander Aircraft Company at the Bethany, Oklahoma, factory. Gulfstream also sold twin-engined Aero Commander rights to Precision Aerospace Corporation of Everett, Washington, but no new production appeared. The company provided support for the large active fleet of twin-engined Aero Commanders.

A revived American Champion firm in Rochester, Wisconsin, achieved modest success with new production of derivatives of the old Champion, including the Scout, Super Decathlon, and Citabria Explorer. Deliveries began in 1991, and in 1996 the total for all models reached 53. Bellanca remained marginally active with the Super Viking, effectively producing to order. Yet another Taylorcraft revival was attempted in November 1989, still involving the original 1930s design. That effort ceased in October 1992, apparently ending the Taylorcraft name permanently. Gilbert Taylor, one of the pioneers of light aircraft, died March 29, 1988, at 89.

Maule, possibly the only wholly family-owned firm in the industry, maintained some 100 employees and output of 60–80 aircraft per year, and remained competitive with its MX-7 series. The Lunar Rocket was discontinued, but later models found steady demand. Many were equipped with floats. Belford Maule died September 2, 1995, aged 83, but the Maule family retained control of the company. The new Maule Comet, with fixed tricycle gear, was offered in the $75,000–$90,000 price range and bid to gain market share. Total production by 1996 reached 2,000 aircraft.

The Ayres firm, also in Georgia and with about 500 employees, continued production of agricultural aircraft and various aircraft and helicopter parts. The basic Ayres model was in the $300,000 price range, at the higher end of the market, and turbine-powered models were even higher-priced. A major development in 1996 was its announcement of the LM 200 Loadmaster light cargo aircraft, together with an order from Federal Express for delivery starting in 1999. An ungainly design with a large fuselage and fixed gear, the Loadmaster featured an innovative power arrangement of coupled turboprops driving a single nose-mounted propeller. To transport up to 7,500 pounds of cargo, the Loadmaster would supplement the current Federal Express fleet of Cessna Caravans, capable of 3,000 pounds.[11]

In February 1995 Schweizer, which produced the original Grumman Ag-Cat for more than 30 years, sold rights to the design to a newly formed Ag-Cat Corporation in Missouri, a subsidiary of an aircraft modification center. The new firm planned to resume production with Schweizer providing assistance during the transition. That venture faded in 1997, however. Air Tractor continued as a strong presence in larger agricultural models, including turboprop versions priced up to $500,000.

Business Jet Survey

Business jets remained a basically sound segment of the general aviation field, although they continued to be widely viewed as a luxury.[12] The business jet segment also was unquestionably the most dynamic and possibly the most competitive of general aviation. There were more firms competing in business jets than in other segments of general aviation aircraft. Major oil companies remained among the heaviest users of business jets, having numerous remote locations around the world that required visits by executives and engineering teams. But by 1991 the business jet market again experienced a severe recession, with deliveries sharply reduced from the rate of immediately prior years. Perhaps reflecting overproduction, some 50 percent of the entire American business jet fleet was on the used aircraft market at any given time. Many corporate jets also were advertised for charter and periodically were made available for humanitarian missions. Especially noteworthy was the Corporate Angel Network, founded in January 1982, in which corporate aircraft transported cancer patients for treatment. Specialized firms also ordered business jets for time-share ownership by companies, with fees paid on the basis of actual usage, an idea with potential for expanding demand. Cessna and others began to market business jets on that basis as well. But the business aircraft market, always cyclical, could not anticipate full recovery outside an overall economic recovery.

Presenting a further challenge, the latest tax changes, effective as of January 1991 as part of a federal deficit-reduction package, included the imposition for the first time of a luxury tax on certain aircraft categories. For general aviation, this involved a 10 percent tax on aircraft costing $250,000 or more unless it could be documented that the aircraft was employed 80 percent of the time for business use. Manufacturers and distributors attributed further declines in sales directly to this tax.

As with other market segments, the business jet field was seriously overcrowded. Competition was also strongly international, similar to the much larger commercial airliner sector, but U.S. producers still commanded about half the world market. International entries included the Canadair Challenger, the Astra Jet series that extended from the Rockwell Jet Commander, and British and French designs. Basic designs offered by the American industry included the Gulfstream IV, Sabreliner, Beechjet 400A (from Mitsubishi), the Citation series, and the Learjet series, including the Model 31A, the tip-tanked Model 35A, and the larger Model 60 replacing the Model 55. The oldest remaining business jet, the Sabreliner, had ended production by 1989, but Sabreliner Corporation remained active with modification and product support for the more than 600 civil models in service. The company also won a Service Life Extension Program (SLEP) contract for the Cessna T-37 fleet in August 1989.

Integrated Resources declared bankruptcy on February 13, 1990, and once again Learjet was on the auction block. Loss of the TTTS contract to Beech was a major setback for Learjet, which developed a version of the Model 31 for the order. Learjet felt that its uncertain future was a negative factor in the competition. Gulfstream issued a letter of intent to purchase Learjet for $60 million, a relative bargain, but the letter expired unexercised on February 22, 1990, due in part to the TTTS loss. Then on April 9, 1990, approval was granted in bankruptcy court for the sale of Learjet to Bombardier of Canada for $75 million. A major challenge for Bombardier was the market perception that Learjet's models, despite continuous refinement, were the least advanced of the current business jet field.[13] Learjet pinned major hopes on the new Model 45, sized between the smaller Model 31 and the long-range Model 60. The Model 45 was to replace the Model 35/36, and its more capacious cabin alleviated criticisms of narrow cabin space in previous models.

Bombardier had expanded rapidly to offer business jets in the small, medium, and large categories, as well as regional airliner models. In addition to the Learjet line and the Canadair Challenger, it acquired Short Brothers of Great Britain, the oldest continuously operating aircraft firm in the world. It then acquired de Havilland Canada from Boeing, in partnership with the Province of Ontario. The Challenger, incidentally, had begun life as the LearStar 600, which Bill Lear designed shortly before his death on May 29, 1978. Canadair purchased the design.

Cessna offered several Citation models. The popular straight-winged Citation II returned to production in 1987, but production of the Citation S/II ended in 1989. Its replacement, the further stretched and more powerful Citation V, first flew on August 18, 1987. While continuing the swept-winged Citation VI and VII, Cessna moved farther up the scale with the still larger Citation X, the most advanced of the series yet developed. Featuring improvements in aerodynamics and engines over the Citation VII, the Citation X made its first flight on December 21, 1993. The upgraded Citation V Ultra was offered from mid-1994, and the Citation V Bravo replaced the veteran Citation II from 1996.

Cessna also reentered the low end of the business jet segment with the new CitationJet of 1991, of the same dimensions as the original Citation I but one ton lighter owing to advanced materials and use of computer-aided design and manufacturing (CAD/CAM).[14] Initially priced at $2.5 million, low for contemporary business jets, the CitationJet sold briskly.

The successor venture of Edward J. Swearingen, Swearingen Aircraft, Inc., announced the small SA-30 Gulfjet in October 1986, with production and marketing to be undertaken by Gulfstream through an agreement of October 1988. Production was planned at the old Aero Commander Oklahoma factory, but Gulfstream withdrew in August 1989, announcing that the SA-30 did not fit its long-range plans. Swearingen also had designed the SA-32T light turboprop military trainer. In 1989 he obtained backing from the Jaffe Group investment firm, enabling him to complete the SA-32T and resulting in the redesignation of the SA-30 as the SJ30. The SA-32T faded, but production of the Swearingen-Jaffe SJ30, which began flight tests in 1990, was planned in Delaware. Swearingen stressed his design's superiority over the directly competitive CitationJet, with a price low enough to induce customers to move up from twin-turboprop models. As with the CitationJet, the SJ30 was powered by Williams/Rolls FJ44 engines but employed simpler technology. Jaffe then withdrew, however, leaving Swearingen to seek new backing, which he eventually obtained from Sino Aerospace of Taiwan. A new Sino-Swearingen Aircraft Company was formed, still in San Antonio. Production then was planned for a new factory in Martinsburg, West Virginia.

The highly publicized Beech Starship experienced various production and certification delays, many attributable to the composite structure, and the first customer deliveries were not made until 1991. The design then encountered market resistance because of its $5 million price tag, equal to that of smaller jets, and a speed which turned out to be significantly lower. Increased weight dictated by certification requirements did not help performance, and reaching the market during a serious recession in executive aircraft was another complication. Slow sales and the rumored breakeven point of 500 aircraft raised widespread skepticism about the Starship's future. Despite criticism that it had simply misjudged the market, Beech maintained that it fully supported the program.[15] But with sales stalled, Beech finally announced in December 1994 that it would cease Starship production, although holding open the option of revival if the market should recover. Only 53 had been built. The King Air series remained strong, however, and Beech planned to continue it indefinitely.

The general aviation industry had been criticized for the lack of a technological revolution, but when developing a model as advanced as the Starship, it encountered market failure. The Piaggio P.180 from Italy, with a similar configuration to the Starship, although of more conventional construction, also became a poor seller. Gates Learjet had invested some $40 million in the P.180 program during 1983–1986 but then withdrew with pessimism about market prospects.

Beech had never developed an original business jet, but its parent, Raytheon, purchased the business jet unit of British Aerospace Corporation for $372 million in August 1993. The sale resulted from a BAe restructuring and consolidation similar in scope to that within the American aerospace/defense industry. The history of product line transfers between firms continued. Raytheon thereby gained a broader market presence with the BAe 125-800 and BAe 1000, larger than the Beechjet and marketed in North America as the Hawker. Raytheon also lessened its vulnerability to the declining military market, while benefiting from the Beech experience with the British design in the 1970s.

The acquisition initially became a new subsidiary, Raytheon Corporate Jets, but on September 15, 1994, Raytheon announced the combination of Beech and Corporate Jets into Raytheon Aircraft Company, headquartered at Wichita. Arthur Wegner became chairman and CEO of Raytheon Aircraft. Production of the jets was sharply curtailed due to market weakness, but was planned to resume at Wichita in 1997. Raytheon then moved to integrate the Beech and Hawker units as business prospects appeared to improve.[16] The Beechcraft nameplate began to fade after more than 60 years. Raytheon Aircraft enjoyed a record year in 1995, with sales exceeding $2 billion. Employment expanded to over 10,000, and a priority was to modernize production facilities, many of which dated from the Second World War.

There was increasing emphasis in business jet development on extended range, with Gulfstream, Bombardier (Canadair), and Dassault proposing models of 6,000–8,000 mile range capability. But the even more ambitious and highly speculative Gulfstream joint venture with the Russian Sukhoi design bureau for a supersonic business jet ended in 1992.

Gulfstream eventually experienced declining sales, which coincided with heavy development expenses for the larger and more advanced Gulfstream V. This led Forstmann Little to intervene in the firm's management. There was widespread skepticism by investors about the prospects for such a large and expensive model, and an attempted public stock offering in 1992 was withdrawn due to lack of support. Forstmann Little then bought out Allen Paulson in September 1992, with Paulson continuing to manage the company. But continued financial deterioration led to Paulson's reluctant retirement in October 1993, although he remained on the board. Theodore Forstmann, while primarily an investor, took the office of chairman and began to assemble a new management team. His firm invested a further $250 million to develop the Gulfstream V, referred to as the GV, while concurrently pursuing cost-cutting and downsizing. The Gulfstream IV, with production past the 250 mark, continued alongside the Gulfstream V.

While a major risk, with a $35 million price and an aggressive and direct competitor in the new Canadair Global Express, the GV soon gained a strong backlog of more than 70 orders. Gulfstream revenues exceeded $1 billion in 1995, operations were profitable, and employment in Savannah exceeded 3,000, marking a complete turnaround after near-bankruptcy.[17] Continued favorable market prospects enabled a successful public offering in 1996 to raise more than $600 million. But a possible future threat was Boeing's announcement in 1996 of plans for an executive version of its Model 737 twin-jet airliner.

By 1995 the business jet segment had improved strongly. New product developments were especially active from that year on. The new Learjet 45 made its first flight on October 7, and deliveries were planned from 1997. The new model made extensive use of CAD/CAM in its development, and with a price in the $8 million range, bid to compete with larger business jet models. The new Gulfstream V also made its appearance with a first flight on November 28, and Raytheon announced its first original business jet, the Premier I. Somewhat larger than the Cessna CitationJet but still competitive with it, the Premier I employed elements of the advanced structure developed for the Starship. The move also indicated a major Raytheon commitment to the business jet segment. The Premier I was expected to be followed by larger models which would eventually succeed the Raytheon/Hawker 1000.

New companies proposed lower-cost single-jet designs. Gaining particular attention was VisionAire, founded in 1988 in Missouri and working in test facilities at Mohave, California. The company developed the Vantage low-cost single-engined business jet, featuring all-composite structure, a forward-swept wing, and design refinements by Bert Rutan of Scaled Composites. Foreign competitors also developed new models.

Business jet marketing returned to an emphasis on cost savings, in part due to escalation in business-class airfares, but principally because corporate downsizing and consolidation required even more travel by hard-pressed executives. Still, business jet use remained a sensitive issue for most corporations: few were interested in publicizing their business jet fleets.

Recovery Prospects

Ironically, the decline and virtual destruction of the general aviation industry occurred during the Reagan Administration, with its strongly pro-business tilt and expressed determination to recapture the spirit of unfettered free enterprise. But general aviation remained a significant economic factor. Throughout the 1980s the size of the U.S. general aviation fleet remained in the 200,000 to 220,000 range, of which more than three-fourths were single-engined fixed-wing types.[18] The average age of single-engined aircraft in 1991 was 26 years.[19] By 1995 the average age of single-engined piston aircraft had increased to 29 years, the average age of turbine-powered business aircraft was 16 years, and for the overall general aviation fleet average age was 26 years.[20] Thus for the longer term, there were bright prospects for new production.

With civil demand depressed, the military market became even more critical to the larger general aviation firms. The business jet field had benefited significantly from military support, with the Learjet, Citation, Beechjet, Gulfstream, and Metro attracting substantial orders. After the Navy ordered the Citation S/II as the T-47A to replace the original North American T-39, the Army selected the new Citation V Ultra, designated the UC-35A, as its high-priority cargo/personnel transport.

General Dynamics, in a reversal of its action of 1985, announced in October 1991 that it would dispose of its Cessna subsidiary. While Cessna remained quite profitable, General Dynamics determined that general aviation did not fit within its new strategic plan, which concentrated on its core technologies. In January 1992 Textron announced the purchase of Cessna for $600 million, which among other things led to a linkage between Cessna and Lycoming engines, also a Textron company. That affiliation was short-lived, however, as Lycoming was sold to AlliedSignal in 1994. Raytheon/Beech had been an interested bidder for Cessna but lost to Textron. There was also the fear of antitrust implications.

The accepted rationale for ownership of general aviation producers by large, financially strong corporations was that it would enable the firms to weather market downturns and maintain research and development programs that might otherwise be terminated. But the problems of Piper under different corporate parents; the frequent ownership changes for such firms as Aerostar, Mooney, Aero Commander, and Bellanca; and the experiences of Cessna with General Dynamics and Gulfstream with Chrysler combined to suggest that ownership of a general aviation manufacturer by a large, diversified parent was not always beneficial to the firm or to the general aviation sector. Furthermore, Beech earlier had been regarded as something of a drag on Raytheon, losing $9 million in 1987, for example.[21]

Both the Beech unit of Raytheon and the Cessna unit of Textron became strong contenders for the important Joint Primary Aircraft Training System (JPATS) program of the Air Force and Navy for their new-generation trainer. The potential $6 billion contract attracted Beech at the beginning, and also drew strong competition from the large military firms, most in collaboration with foreign firms. Then after lengthy delays in the selection process, Cessna entered a new design based on the wing and power of the CitationJet.

The eventual winner, the turboprop Beech Mark II, was developed in cooperation with Pilatus of Switzerland from its established PC-9. It received the military designation T-6A. The contract award, announced in June 1995, sparked vigorous protests from Textron, and was a major victory for Raytheon. The combination of TTTS/T-1A production extending into 1997 and JPATS/T-6A orders would enable Raytheon/Beech to enjoy the cushion of military production, along with strong export prospects for the trainer, while awaiting a civil market recovery.

The most important step toward that recovery, long sought by the industry, was product liability reform. Legislative reform finally was realized with passage of the General Aviation Revitalization Act (GARA), signed into law by President Clinton on August 17, 1994. The law provided for an 18-year Statute of Repose, meaning that new aircraft are not subject to product liability claims after 18 years. Particularly encouraging for the industry was that by 1997–1998, the potential for new product liability lawsuits would drop significantly, since aircraft of the high-production years of the late 1970s would then be excluded. Critics of the legislation maintained that the threat of product liability lawsuits led to more safety advances and innovation than would have been the case otherwise, but the law was hailed by the industry.

Russ Meyer announced immediately that Cessna would reestablish production of the veteran single-engined Models 172, 182, and 206, forecasting an output of some 2,000 annually. A new plant in Independence, Kansas, was to be built. Cessna was expected to expand its offerings with market recovery, but such plans likely would not include revival of the piston-engined Golden Eagle and Chancellor, as turbine power was strongly favored in that segment. Nor would they include revival of the Skywagon, given the success of the Caravan.[22] With improving prospects for Cessna, Raytheon/Beech, and Learjet, the city of Wichita, which had suffered with the industry contraction, appeared to be on the verge of a long-anticipated recovery.

Building on the recovery of the oil service market, the market for larger general aviation helicopters also was showing signs of vitality by 1995. Executive use of helicopters increased in a strengthening economy, and the international market was also a strong growth area. But civil helicopters remained the smallest segment of the overall aerospace industry. Production volume, while increasing, remained low. Helicopter use was still limited by cost, with the leading small turbine helicopter, the Bell Model 206B JetRanger III, carrying a base price of some $750,000. Choice was widespread, however, with Bell, McDonnell Douglas, and several smaller manufacturers as factors. The improved Sikorsky S-76B, which appeared in 1988, was primarily for the export market, but a few executive models were ordered.

Among smaller firms, Enstrom production remained sporadic; eleven were delivered from the Menominee, Michigan, factory in 1995. But in 1994 the company implemented an agreement for helicopter production in China, possibly enhancing its future. Robinson gained orders for 179 aircraft in 1995, encouraging given a base price of $135,000 for the two-seat model. Stanley Hiller's son Jeffrey repurchased the Rogerson-Hiller rights in 1993 with a view toward reviving production after several years of inactivity. Again relocated, this time to Newark, California, and backed by Thai investors, Hiller restarted production of the veteran UH-12E3. The first new production model first flew on June 2, 1995. Schweizer delivered 29 helicopters in 1995.

The I … was succeeded by the greatly advanced Model
430, … ded rotor. It first flew on October 25, 1994. The
Mod … ajor stretch, with capacity for eight passengers,
and … ared promising despite being priced in the $4
milli … uction deliveries began in June 1996. The new,
sma … Model 407, optimized for business use and
pric … .3 million, quickly gained strong orders, with
85 … g 1996. To succeed both the JetRanger and
Lon … ill powered by the Allison 250, the Model 407
feat … ded rotor and a larger, roomier cabin.[23] All were
pro … BHC subsidiary, which continued as a strong
suc … 1500 BHC helicopters had been built through
199 … commercial helicopters in that year alone.

In … in the civil market, Kaman, the long-established
mil … producer, developed the unusual K-MAX, which
firs … mber 23, 1991. A single-seater with a narrow
fus … nation skid-and-wheel landing gear, the K-MAX
wa … carry external cargo as an "aerial truck." The
K- … d an advanced intermeshing rotor blade system
an … by a 1,500 shp Lycoming turboshaft. First de-
liv … 94. Priced in the $3.5 million range, the K-MAX
er … ng domestic and export market prospects for a
va … l applications, especially logging.

T … ouglas Helicopter Company was renamed McDon-
n … copter Systems in September 1993. The stretched
M … ive with the Bell Model 430, was certificated for
s … . The most advanced model, the eight-passenger,
t … Explorer (originally MD 900), was launched in
1 … iveries at the end of 1994. A multinational NOTAR
c … ed in the $3 million range, The MD Explorer
f … rticipation by Australia and Japan. The MD500, at
s … e more powerful MD530, and the MD520 NOTAR
c

… otential downside to the recovering general avia-
… ng sector was that the kitplane segment, enjoying

government support and increasing public acceptance, could be harmed by revitalized production of smaller single-engined factory models. Kit sales (see Table 8-1) already had leveled off in the 1990–1993 period after rapid growth during the 1980s. Kitplane construction had circumvented product liability lawsuits, but with reform, factories could again produce aircraft that challenged the homebuilts. The revived Cessna 172R, for example, carried a base price of $124,500, making it price-competitive with four-seat kit designs. Product development remained highly active, however, aided by the new Small Aircraft Certification Compliance Program of the FAA. The advanced Cirrus Design SR20 and Stoddard-Hamilton GlaStar attracted particular attention.[24]

Internationalization of the general aviation sector increased rapidly, although products of most producing nations maintained a clear national identity. Marketing had long been international, but development and production were becoming increasingly globalized. The activities of Bell in helicopters, Bombardier in business jets and commuter aircraft, and widespread component sourcing between the two nations blurred the division between the U.S. and Canadian industries. Integration was increasing. The new Learjet Model 45 in particular benefited from its corporate linkages with Canadair and

Table 8-1
Kit and Plans Aircraft Sales

Year	Kits	% Change	Plans	% Change
1990	1,147	N/A	2,560	N/A
1991	1,573	37	2,209	−14
1992	1,943	24	1,753	−21
1993	3,082	59	2,116	21
1994	4,085	33	2,831	34
1995	5,338	31	3,063	8
1996	5,713	7	2,645	−14
1997*	5,750	1	3,205	21
1998*	6,921	20	1,761	−45
1999*	7,751	12	1,606	−9

* Manufacturer's projections.
SOURCE: *Kitplanes,* July 1997, p. 33.

Short Brothers in its development, and had major components built in Canada and Northern Ireland.

In another manifestation of internationalization, Fairchild Aircraft acquired 80 percent of the German firm Dornier Luftfahrt GmbH on June 5, 1996. Fairchild Dornier planned to develop the German Models 226 and 326 commuters, larger than the 19-seat Metro II, to offer broader coverage of the commuter market. Daimler-Benz, parent of Dornier, had wanted to dispose of the company.

The agenda for the future of general aviation was encompassed within the term revitalization. Product liability reform had been the chief priority, but with success on that issue, attention turned to such matters as flight training, international harmonization of rules and standards, increased funding and greater autonomy for the FAA, and greater efficiency in certification matters.[25] By 1997 there were strong indications that the revitalization program was showing results. The number and variety of new general aviation aircraft appearing was encouraging, and GAMA reported a record $3.1 billion in 1996 billings. The figure included fixed-winged aircraft only. Its member firms reported delivery of 1,132 aircraft for 1996, led by Raytheon, Cessna, Gulfstream, and Learjet.[26]

Perhaps portending a substantive response to the criticism of static technology in general aviation, NASA, already active in research on fuel-efficient engines and advanced propeller blades, announced that it would assist in development of new advanced-technology small aircraft engines. The Advanced General Transport Experiments (AGATA) program, a joint initiative of NASA, the FAA, universities, and industry, undertook long-range development of advanced technology for fixed-wing light aircraft. There were also indications that future single-engined aircraft would make greater use of composites in their structures.

# # # # #

The general aviation industry has experienced a turbulent history. Long-term success has been elusive, with bankruptcies all too common. If any single conclusion emerges, it is that attaining consistent profitability in general aviation aircraft has been exceptionally difficult. The three facets of the business, development, production, and marketing, must be combined for success, and many promising designs have failed due to production and financing problems.

General aviation has been further handicapped by its erratic growth and financial instability. Potential customers almost inevitably are inhibited if they doubt the ability of the manufacturer to survive. Furthermore, although having grown over time, general aviation has never attained the status of a "hot" industry, and never attained the degree of political influence of the much larger military, commercial, and space sectors of the aerospace industry. It is unquestionably the least appreciated, most misunderstood sector of aviation. The public still does not recognize the importance of business flying. Yet the value and importance to the economy of general aviation are beyond question.

The production volume of the 1970s for general aviation aircraft is unlikely to be regained, but the existing general aviation fleet eventually will need replacement. Recent legal and regulatory reforms have been important, if not essential, but perhaps more important are the eventual replacement requirement for the existing fleet and continuation of the entrepreneurial spirit which has characterized the field.

In favor of the industry's future is that it has largely carried out a major, fundamental restructuring. While forced by a combination of legal and economic factors, that restructuring is now largely complete and should result in an overall stronger industry sector, albeit with fewer producers. That restructuring, combined with economic expansion, expanded international markets, and continued technological development, provides a basis for guarded optimism for the future of the industry and for the growth of general aviation.

Appendix

General Aviation Industry Chronology, 1919–1997

1920 1930 1940

Weaver Aircraft Co. (1919)
(Successor firm to Lorain, OH, 1921)
(to) Advance Aircraft (1923)
(to Troy, OH, 1923)
To WACO (1929)

Wichita Aircraft (1919)
(to) E. M. Laird Co. (1920)
(to) Swallow Airplane (1924)
(reorg. Swallow Airplane Co., 1927)—
(exit, 1931)///

(new) E. M. Laird Co. (1926)

Laird Aircraft Corp. (1927)-(exit, 1930)///

Travel Air Corp. (1924)
(to Curtiss-Wright, 1929)—(TA exit, 1933)///

Curtiss-Wright Corp. (1929)
(exit personal aircraft, 1937)

Beech Aircraft (1932)

Robertson Aircraft (1923)

(new) Curtiss-Robertson Div. C-W Corp. (1929)
(absorbed 1933)

Stearman Aircraft, Inc. (1926)
Stearman Aircraft Corp. (1927)
(Div. United A&T, 1929)
(Stearman Div. Boeing, 1934)

Stinson Airplane (1920)
Stinson Airplane Syndicate (1925)
Stinson Aircraft Corp. (1926)
(Div. Cord Corp., 1929)
(Div. Aviation Corp., 1933)
(Div. Aviation Mfg. Co. AVCO, 1934)

Barkley-Grow (1936)

Buhl-Verville (1925)
Verville Aircraft (1927)—(exit, 1931)///

Alexander Aircraft (1925)——(exit, 1932)///
Aircraft Mechanics (1932)
(exit, 1934)////

Cox-Klemin (1921)—(exit, 1925)///

Fairchild Aviation (1925)
Fairchild Airplane Mfg. (1927)-(to AVCO, 1929)

Kreider-Reisner (1925)
(to Fairchild as Division, 1929)

(Fairchild withdraws, 1931)

(new) Fairchild Aircraft (1934)
(reorg.) Fairchild Engine & Airplane Corp. (1936)

G. Elias & Bro. (1919)—Elias Aircraft (1929)
(exit, 1931)////

Gates-Day Aircraft (1927)
(new) Standard Aircraft (1928)—(exit, 1931)//

American Eagle (1926)
(affl. Lincoln-Page, 1931)
(reorg. American Eagle-Lincoln, 1931)
(exit, 1933)////

Porterfield Aircraft (1934)

Bellanca-Roos (1922)
(exit, 1924)///

Bellanca Aircraft (1927)

1920 1930 1940

Central States Aero (1926)
Mono Aircraft (1929)
(exit, 1931)////

(rights to) Monocoupe Corp. (1931)
(reorg.) Lambert Aircraft (1933)
(rights to) Dart Mfg. Corp. (1937)
(rights to) Culver Aircraft (1939)-

(new) Monocoupe Corp. (1937

Luscombe Airplane Dev. (1935)
Luscombe Airplane (1938)

Cessna-Roos (1927)
Cessna Aircraft (1927)
(production suspended, 1931–34)

Rearwin Airplanes (1929)
Rearwin Aircraft & Engines (1937)

Howard Aircraft Corp. (1937)

Bennett Aircraft (1939)
Globe Aircraft (1941)

Mid-Continent Aircraft (1928)
(renamed) Spartan Aircraft (1928)

Cunningham-Hall (1928)

Pitcairn Aircraft (1927)
(subs. Pitcairn Aviation)
Pitcairn Aircraft (1928)
(Div. Pitcairn-Cierva Autogiro Co., 1929)
Autogiro Corp. of America (1931)
(to) Pitcairn-Larsen(1940)

Heath Aircraft (1926)—
(exit, 1931)///

Aeronautical Corp. of America (1928)

Taylor Brothers (1927)
Taylor Aircraft (1931)—
(reorg.) Piper Aircraft (1937)

(new) Taylor Aircraft (1936)
Taylor-Young (1937)
Taylorcraft Aviation (1939)

Meyers Aircraft (1936)

Call Aircraft (1939)

Ryan Aeronautical (1934)

Engineering and Research Corp. (1930)
(enters aircraft, 1936)

Grumman (1929)
(enters personal aircraft, 1937)

Interstate Aircraft & Engineering (1937)

Harlow Aircraft (1939)

1940 1950 1960

Waco—(exit, 1947)///

Beech—

Cessna—

Robertson—
(exit, 1946)///

Stinson Div.
Convair—(rights to
Piper, 1948)
(production suspended, 1950)////

Barkley-Grow—
(to Vultee, 1940, in
Vultee-Stinson Div.)///

Fairchild E & A—(exit private —
aircraft, 1947)

Porterfield
(exit, 1942)///

Bellanca—(production
suspended, 1952)/////
(rights to
Northern Aircraft, 1955)—

Aeronca Corp—(exit aircraft, 1951)///
(renamed, 1941) (rights to
Champion Aircraft, 1954)-

Monocoupe—(exit, 1947)///

Culver—(exit, 1946)////////(rights to
Superior Aircraft, 1956)
(exit, 1959)////

Luscombe—Bankrupt, 1948
(reorg., 1949)
(to Temco, 1949)
(production suspended, 1950,
absorbed, 1953)
(rights to Silvaire Aircraft, 1955)—

Texas Engr.
and Mfg. Co. (1946)
(renamed Temco, 1946)—
(reorg.) Temco
Aircraft Corp., 1952—
(exit personal
aircraft, 1950)///

Globe—bankrupt, 1946///
(rights to Temco, 1947)

Rearwin
(to Commonwealth, 1942)
(exit, 1947)///

Interstate
(to Harlow, 1945)
Harlow—(exit, 1946)///

Spartan—(exit, 1947)///

Cunningham-Hall—(exit, 1948)///

ERCO—(exit, 1950)
(rights to Fornaire, 1955)—
(rights to
Carlsbad, N. M., 1960)—

Howard
(exit, 1944)///

Pitcairn-Larsen
(to AGA Corp., 1941)
(to Firestone as G&A, 1942)--(exit, 1947)///

Piper—
(production suspended, 1947)

Taylorcraft—bankrupt, 1946
|
(new) Taylorcraft Corp., 1947—(bankrupt,1959)

Meyers—
(production suspended, 1956)—(revived, 1959)

Call—
(incorp. CallAir, 1959)

Republic—
(exit personal aircraft, 1947)///

Ryan—
(exit personal aircraft, 1951)///

1940 1950 1960

North American—
(exit personal aircraft, 1947)///

Koppen-Bollinger (1948)
(renamed Helio, 1949)—

Mooney Aircraft (1948)—
(reorg. 1954)

Aero Design and Engr. (1944)—(to Rockwell-
(reorg. 1950) Standard, 1958)

Colonial Aircraft (1946)—
(rights to Lake Aircraft,1959)—

Anderson-Greenwood (1941)0
(exit aircraft, 1948)//////

Fletcher (1941)—
(reorg. Flair, 1960)

Baumann (1945)—(exit, 1952)///

Custer
Channel Wing (1956)—

Hiller (1942)
(renamed United Helicopters 1945)
(renamed Hiller Helicopters 1950)—

Brantly (1943)—

Siebel (1948)—
|
(to) Cessna, 1952—

Enstrom (1959)-

Bell—
(enters civil
helicopters, 1946)

B. F. Maule Co.(1941)—(reorg. Maule—
Aircraft, 1961)

Snow (1958)—

Grumman—
(reenters general
aviation, 1957)

1960 1970 1980

Swiss-American (1960)
|
Lear Jet Industries (1962)—
(to) Gates Rubber (1967)
(reorg. Gates Learjet, 1970)—

Brantly—(to Lear Jet, 1966)
(sold to ARDC, 1969)
|
(to Brantly-Hynes, 1975)—

Enstrom—(to Purex, 1968)
(production suspended, 1970)
(to F. Lee Bailey, 1971)—

Robinson (1973)—

Beech—
(to) Raytheon Corp.(1980)

Cessna—

American Aviation (1964)—
(to) Grumman Corp. (1973)—
(as GAAC)
|
(to AJI, Inc., 1978)—

Grumman—
(exit general aviation, 1979)

Flair—
(1960) |
(reorg. Sargent-Fletcher (1964)
(to AJI, Inc., 1964)
(exit aircraft, 1964)////

Custer—(exit, 1968)/////

Swearingen—
(enters aircraft production 1965)
|
(to) Fairchild (1971)—

Hiller
(to ELTRA, Inc., 1960)—(to Fairchild, 1964)—
|
(Fairchild sells, 1973)—

Northern Aircraft
(to) Downer (1959)
|
(to) Inter-Air, 1964
(to) Miller, 1966—
(reorg.) Bellanca Aircraft (1970)
—(to) Anderson-
Greenwood, 1976—

Champion (to Bellanca, 1970)

(new) Bellanca, 1960—(exit, 1975)////

Silvaire (exit, 1962)/////

Helio—(acq. by General Aircraft, 1969)
(exit, 1976)////

(new) Alon, Inc. (1963)
(Ercoupe rights)
|
(to Mooney, 1967)

Mooney—(to A. E. L., Inc., 1969)
(rights to Butler, 1969)
Ted Smith Aircraft (1963)—(to Butler Int'l, 1970)
|
(Mooney & Ted Smith combined as
(Aerostar Corp., subsidiary
Butler Int'l, 1970)—
(production suspended, 1972)

(new) Ted Smith & Assoc.(1972)
|
(to) Piper, 1978—

Mooney (to) Republic
Steel, 1973—

Piper—
(acq. by Bangor Punta, 1969)

(new) Taylorcraft (1968)—

CallAir
(bankrupt, 1961)
(to IMCO, Inc., 1962)
(to R-S Corp., 1965)

Volaircraft (1963)
(to) R-S Corp., 1965

1960 1970 1980

Snow—(to) R-S Corp., 1966
(incorp., 1961)

Meyers—(to) R-S Corp., 1965

Rockwell Aero Commander (1965)—
(subs. NAR, 1967)

Lake—(to Consolidated
Aeronautics, 1962)—

(new) Navion
Aircraft (1960)
(rights from Ryan)
|
Navion Corp., 1965—(to) Navion Rangemaster (1972)
(exit, 1975)/////

Air Tractor, Inc. (1971)—

Ayres (1977)—
(rights from Rockwell
International)

Bell—

1980 **1990** **1997**

Gates Learjet
(to) Integrated Resources (1987)
(to) Bombardier (1990)

Beech
(to) Raytheon (1980)
Raytheon acq. BAe Business jets, 1992
(merged)Raytheon Aircraft (1994)

Rockwell Aero
Commander
(to) Gulfstream
American
(acq. RI Gen. Aviation
interests, 1981)
(production suspended, 1982)
(name chg. Gulfstream Aerospace, 1982)
(to) Chrysler (1985)
(to) Forstmann Little (1989)

American General (1989)—Bankrupt, 1994///
(former Grumman designs)

Fairchild
Aircraft Corp. (1982)—(to)
(former Swearingen) GMF Corp., 1987—
(to) new investors (1990)

(new) Swearingen Corp. (1982)-(to) Sino-
Swearingen (1995)

Maule—(bankrupt, 1984)
(reorg. Maule Air, Inc, 1984)

Bellanca
(sold by A-G, production suspended, 1981)
(new) Bellanca, Inc. (1982)
(Champion production suspended, 1982)

(new) Champion (1982)—(exit, 1985)///

(new) American Champion Corp. (1991)

Taylorcraft (bankrupt, 1986)///

Cessna—(acq. by General Dynamics (1985)
(acq. from GD by Textron, 1992)

Mooney—(reorg.) Mooney Aircraft
Corp. (1984)

Piper—(acq. by Millar, 1987)
(parent Bangor Punta
acq. by Lear-Siegler, 1984)
(bankrupt, 1991)
(reorg. New Piper
Aircraft, Inc., 1995)

Ayres

Lake Amphibians
(reorg., new owners, 1981)

Air Tractor, Inc.

Hynes (1984)—(exit, 1987)///

Enstrom (to Bravo, 1980)
(to new investors, 1984)

(new) Hiller (1980)—(bankrupt, 1984)
(to) Rogerson-Hiller (1984)
(production suspended, 1984)
(rights to new Hiller, 1993)

Bell
(est. Bell Helicopter
Canada, 1982)

Robinson

Aviat (1979)

Stoddard-Hamilton (1979)

Cirrus Design (1986)

Notes

Chapter 1

1. Tom Crouch, "General Aviation: The Search for a Market," in E. M. Emme (ed.), *200 Years of Flight in America: A Bicentennial Survey,* Univelt, San Diego, 1977, p. 113.
2. Alexander Klemin, "Investing in Aviation," *Scientific American,* February 1929, p. 149.
3. Roger Bilstein, *Flight Patterns,* University of Georgia Press, Athens, GA, 1983, p. 67.
4. Herm Schreiner, "The Waco Story. Part I: Clayton Brukner & the Founding Years," *American Aviation Historical Society Journal,* vol. 25, Winter 1980, p. 282.
5. John T. Nevill, "The Story of Wichita," *Aviation,* November 1930, p. 291.
6. Ibid., December 1930, p. 355; and Frank Joseph Rowe and Craig Miner, *Borne on the South Wind,* Wichita Eagle and Beacon Publishing Co., 1994, p. 67.
7. Edward H. Phillips, *Cessna: A Master's Expression,* Flying Books, Eagan, MN, 1985, p. 35.
8. Edward H. Phillips, *Travel Air: Wings over the Prairie,* Flying Books, Eagan, MN, 1982, p. 7.
9. William H. McDaniel, *The History of Beech,* Beech Aircraft Corp., Wichita, 1982, p. 12.
10. Rowe and Miner, op. cit., p. 97.
11. Mitch Mayborn and Peter M. Bowers, "A History of the Stearman Aircraft Company," in *Stearman Guidebook,* Flying Enterprise Publications, Dallas, 1973, pp. 4–5.
12. Phillips, *Travel Air,* p. 44.
13. Paul R. Matt, *Alfred Victor Verville,* vol. 18: *Historical Aviation Album,* Historical Aviation Album, Corona Del Mar, CA, 1987, pp. 93–94.
14. John W. Underwood, *The Stinsons,* The Heritage Press, Glendale, CA, 1969, p. 29.
15. *Aircraft Year Book, 1929,* Manufacturers Aircraft Association, New York, p. 74.

16. "Fairchild Company Buys Control of Kreider-Reisner," *Wings of Industry,* April, 15, 1929, p. 1.
17. *Jane's All the World's Aircraft, 1929,* Sampson Low, Marston and Co., London, p. 269; and *Aircraft Year Book,* 1929, p. 75.
18. "Cessna Airplane Factory is Located at Wichita," *Aviation and Aeronautical Engineering,* December 1, 1916, p. 294.
19. Phillips, op. cit., pp. 32–33.
20. Ibid., p. 32–33 and 38; and Nevill, op. cit., December 1930, p. 354.
21. David B. Stevenson, "Charles Healy Day and His New Standards," *American Aviation Historical Society Journal,* vol. 41, Fall 1996, p. 207.

Chapter 2

1. Grover Loening, *Takeoff into Greatness: How American Aviation Grew so Big so Fast,* Putnam, New York, 1968, p. 172.
2. Crouch, op. cit., pp. 122–123.
3. Bilstein, op cit., pp. 68–71.
4. *Aerospace Facts and Figures, 1959,* Aerospace Industries Association, Washington, D.C., pp. 6–7.
5. Richard P. Hallion, *Designers and Test Pilots: The Epic of Flight* series, Time, Inc., New York, 1983, pp. 40–41.
6. Robert H. Rankin, "The Genius of Giuseppe," *Flying,* February 1953, pp. 36–37 and 60–61.
7. Hallion, op. cit., pp. 42–45.
8. Rankin, p. 61.
9. *Jane's All the World's Aircraft, 1929.*
10. Paul R. Matt, *Aeronca: Its Formation and First Aircraft, Historical Aviation Album,* vol. 10, Historical Aviation Album, Corona del Mar, CA, 1971, p. 227.
11. Bilstein, op. cit., p. 70.
12. Matt, *Aeronca,* op. cit.
13. Devon Francis, *Mr. Piper and His Cubs,* Iowa State University Press, Ames, IA, 1973, pp. 16–17, 22, and 41.
14. William F. Trimble, *High Frontier: A History of Aeronautics in Pennsylvania,* University of Pittsburgh Press, Pittsburgh, 1982, p. 183; and "Count the Cubs," *Fortune,* June 1940, p. 113.

15. Randolph F. Hall, "Cunningham-Hall Aircraft Corp. Story," *American Aviation Historical Society Journal,* vol. 16, Summer 1971, p. 90.
16. Trimble, op. cit., p. 174.
17. Frank Kingston Smith, *Legacy of Wings: The Story of Harold F. Pitcairn,* J. Aronson, New York, 1981, pp. 26 and 29.
18. Trimble, op. cit.
19. James J. Horgan, *City of Flight: The History of Aviation in St. Louis,* The Patrice Press, Gerald, MO, 1984, p. 302.
20. Frederick W. Roos, "Curtiss-Wright St. Louis," *American Aviation Historical Society Journal,* vol. 35, Winter 1990, p. 305.
21. McDaniel, op. cit., p. 9.
22. Roos, op. cit., p. 293.
23. Phillips, *Cessna.*
24. McDaniel, op. cit.
25. *Aviation,* May 1933, p. 148.
26. Genevieve Brown, *Development of Transport Airplanes and Air Transport Equipment,* Air Technical Service Command, Wright Field, OH, April 1946, p. 67.
27. Horgan, op. cit., p. 321.
28. *Jane's All the World's Aircraft, 1942,* p. 209.
29. Bob Whittier, "Light Plane Heritage," *EAA Experimenter,* December 1993, p. 31–32.
30. Eugene L. Vidal, "Low-priced Airplane," *Aviation,* February 1934, pp. 40–41.
31. Joseph J. Corn, *The Winged Gospel: America's Romance with Aviation,* Oxford University Press, New York, 1983, pp. 98–100.
32. William F. Trimble, "The Collapse of a Dream: Lightplane Ownership and General Aviation in the United States after World War II," in William F. Trimble (ed.), *Pioneers and Operations,* vol. 2: *From Airships to Airbus: The History of Civil and Commercial Aviation,* Smithsonian Institution Press, Washington, D.C., 1995, p. 129.
33. Crouch, op. cit., p. 117; and Roger Bilstein, *Flight in America,* Johns Hopkins University Press, Baltimore, 1984, p. 109.

Chapter 3

1. Trimble, *High Frontier,* p. 185. "Count the Cubs," *Fortune,* June 1940, gives a figure of 523 for that year.

2. Ibid.
3. Chet Peek, *Taylorcraft: The Taylorcraft Story,* SunShine House, Terre Haute, IN, 1992, p. 24.
4. Ibid., p. 43; and Donald Ross, *An Appraisal of Prospects for the Aircraft Manufacturing Industry,* White, Weld and Co., New York, 1940.
5. Francis, op. cit., pp. 46–59.
6. William G. Cunningham, *The American Aircraft Industry: A Study in Industrial Location,* Lorrin L. Morrison, Los Angeles, 1951, p. 68.
7. John C. Swick, *The Luscombe Story: Every Cloud Has a Silvaire Lining,* SunShine House, Terre Haute, IN, 1987, pp. 65–68.
8. Smith, Frank K., op. cit., p. 301.
9. Trimble, *High Frontier,* p. 219.
10. McDaniel, op. cit., pp. 24–28.
11. Phillips, *Cessna,* pp. 99–104.
12. "Stubborn Cessna toiled day, night," *Aviation Pioneers,* Wichita Eagle-Beacon Publishing Co., Wichita, 1987, p. 5.
13. Kenneth D. Wilson and Thomas E. Lowe, "Lloyd C. Stearman, 1898–1975," *American Aviation Historical Society Journal,* vol. 36, Summer 1991, p. 90.
14. Crouch, op. cit., p. 123.
15. Underwood, op. cit., pp. 72–74.
16. Almarin Phillips, *Technology and Market Structure: A Study of the Aircraft Industry,* Heath Lexington, Lexington, MA, 1971, p. 92.
17. Bill Gunston, *One of a Kind: The Story of Grumman,* Grumman Corp., Bethpage, NY, 1988, p. 24.
18. *Of Men and Stars: A History of Lockheed Aircraft Corporation, 1913–1957,* Chapter 5, Lockheed Aircraft Corp., Burbank, CA, 1957, p. 8.
19. James R. Wilburn, "Social and Economic Aspects of the Aircraft Industry in Metropolitan Los Angeles during World War II," unpublished Ph.D. dissertation, UCLA, 1971, pp. 11–12; and letter, Robert E. Gross to Courtlandt S. Gross, March 31, 1937, Gross Papers, Box 13, Library of Congress.
20. Carl J. Peterson, *The CallAir Affair,* privately published by the author, 1989, p. 16.
21. Cunningham, op. cit., p. 175.
22. William Wagner, *Ryan, the Aviator,* McGraw-Hill, New York, 1971, p. 178.

23. Grover Loening, *Aviation and Banking,* Chase National Bank, New York, October 1938, p. 10.
24. Alfred Goldberg (ed.), *History of the United States Air Force,* Van Nostrand Co., Princeton, NJ, 1974, p. 118.
25. Francis, op. cit., pp. 75 and 82.
26. Underwood, op. cit. pp. 66–67.
27. "Beechcraft Takes Off on the Wings of a Jet," *Business Week,* February 25, 1956, p. 180.
28. McDaniel, op. cit., pp. 51–58.
29. Phillips, *Cessna,* pp. 119–121.
30. Trimble, *High Frontier,* p. 183.
31. Swick, op. cit., p. 89.

Chapter 4

1. Phillips, *Cessna,* p. 121.
2. Lew Townsend, "Lightplane Mergers: Beech, Cessna, Lear Didn't Get Together," *Wichita Eagle-Beacon,* Wichita, February, 10, 1990, pp. 1F and 8F.
3. *Aerospace Facts and Figures, 1959,* p. 7.
4. *Aircraft Facts and Figures, 1957,* Aircraft Industries Association, Washington, D.C., p. 7, states 1,417. Retired General Echols, in his speech to the AIA in 1947, mentions a figure of 1,330 military aircraft produced for 1946, the same figure quoted by Robert E. Gross of Lockheed.
5. George Bryant Woods, *The Aircraft Manufacturing Industry: Present and Future Prospects,* White, Weld, New York, 1946, p. 46.
6. *Aircraft Year Book, 1946,* Aircraft Industries Association, Washington, D.C., p. 235; and Crouch, op. cit., p. 126. Joseph T. Geuting, Jr., manager of the Private Aircraft Council, made such an estimate.
7. Bilstein, *Flight in America,* p. 195.
8. John Foster, Jr., "The Personal Plane Sales Target" (survey for Parks Air College), *Aviation,* January 1944, pp. 116–117.
9. *NACA-Industry Conference on Personal Aircraft Research,* Langley Memorial Aeronautical Laboratories, Langley Field, VA, September 20, 1946.
10. Woods, op. cit.

11. Edward T. Austin, *Rohr: The Story of a Corporation,* Rohr Corp., Chula Vista, CA, 1969, pp. 41–42.
12. Richard Thruelson, *The Grumman Story,* Praeger, New York, 1976, pp. 225–226.
13. Herbert Solow, "North American: A Corporation Deeply Committed," *Fortune,* June 1962, p. 164.
14. Wagner, op. cit., p. 226.
15. "Experiment at Republic," *Fortune,* February 1947, p. 167.
16. "New Planes for Personal Flying," *Fortune,* February 1946, p. 126.
17. *General Aviation Statistical Databook, 1990–1991,* GAMA, Washington, D.C., 1992, p. 4.
18. Peek, op. cit., p. 167.
19. Trimble, *High Frontier,* p. 244.
20. Hall, op. cit., p. 97.
21. Swick, op. cit., pp. 106–107 and 152.
22. Rowe and Miner, op. cit., pp. 159–160.
23. McDaniel, op. cit., pp. 67 and 118; and Frank E. Hedrick, *Pageantry of Flight: The Story of Beech Aircraft Corporation,* Newcomen Society, New York, September 28, 1967, p. 23.
24. Trimble, *High Frontier,* p. 243; and Francis, op. cit., pp. 139–141.
25. Francis, ibid., pp. 142–148.
26. Underwood, op. cit., p. 76.
27. Trimble, *High Frontier,* p. 244; and Francis, op. cit., p. 148.
28. Francis, op. cit., pp. 151–161.
29. Thruelson, op. cit., p. 226.
30. Gerard P. Moran, *Aeroplanes Vought, 1917–1977,* Historical Aviation Album, Temple City, CA, 1978, pp. 130–133.
31. Nicholas M. Williams, "The Aero Commander 520," *American Aviation Historical Society Journal,* vol. 35, Spring 1990, pp. 20–25.
32. Corn, op. cit., p. 140.
33. *Aviation Facts and Figures, 1953,* Aircraft Industries Association, Washington, D.C., pp. 140–141.
34. Cunningham, op. cit., pp. 196–197.

Chapter 5

1. *Aviation Facts and Figures,* 1959, p. 86.

2. George Hardie, Jr., "The Long Road Back: The American Airmen's Association," *Sport Aviation,* June 1986, pp. 35–37.
3. Keith Beveridge, *Kitplanes,* July 1997, p. 33.
4. Gordon Baxter, "Mooney," *Flying,* September 1977, p. 258.
5. Thomas J. Harris, *The Magic of Aero Design,* Newcomen Society, New York, May 10, 1962.
6. Williams, op. cit., p. 37.
7. Petersen, op. cit., pp. 18–23.
8. *Jane's All the World's Aircraft, 1967–68,* p. 272.
9. Francis, op. cit., p. 181.
10. Trimble, *High Frontier,* p. 261.
11. *An Eye to the Sky,* Cessna Aircraft Company, Wichita, 1962, pp. 66 and 262.
12. *Flight International,* vol. 91, June 29, 1967, p. 1074.
13. "Beechcraft Takes Off on the Wings of a Jet," pp. 178–180.
14. Hedrick, op. cit., p. 32.
15. Gunston, op. cit., pp. 74–75.
16. David A. Brown, "Aero Commander Management Shifts Set," *Aviation Week & Space Technology,* July 19, 1965, pp. 20–21.
17. "Aero Commander Plans to Market Full Single-Engine Aircraft Line," *Aviation Week & Space Technology,* March 2, 1964, p. 36.
18. *Aviation Week & Space Technology,* April 13, 1964, p. 28.
19. Crouch, op. cit., p. 126.
20. "500 Largest Industrial Corporations," *Fortune,* June 15, 1968, pp. 188–204.
21. "A Look at America's Newest Volume Industry in 1970," internal study, Cessna Aircraft Company, Kansas Aviation Museum papers, circa 1960.
22. McDaniel, op. cit., p. 135.
23. Robert J. Serling, *The Jet Age: The Epic of Flight* series, Time, Inc., New York, 1982, p. 112; and Richard Rasche, *Stormy Genius,* Houghton Mifflin, Boston, 1985, pp. 223–226.
24. Jeffrey Ethell, *NASA and General Aviation,* NASA, Washington, D.C., 1986, p. 14.
25. Rasche, op. cit., pp. 226 and 253.
26. Ibid., p. 289; and Townsend, op. cit., p. 8F.

Chapter 6

1. Crouch, op. cit., p. 111.

2. Ethell, op. cit., pp. 4 and 11.
3. *General Aviation Statistical Databook, 1990–1991,* p. 4.
4. "Crosswinds," *Forbes,* November 1, 1968, p. 44.
5. "A Divided House at Piper," *Business Week,* March 16, 1974, p. 94.
6. Francis, op. cit., pp. 219–223.
7. *Aviation Week & Space Technology,* April 19, 1970, pp. 67–69; and Trimble, *High Frontier,* p. 283.
8. Erwin J. Bulban, "Piper, Swearingen Study Link; Latter Developing Business Jet," *Aviation Week & Space Technology,* July 26, 1971, p. 19.
9. "A Divided House at Piper," p. 94.
10. Richard J. Levine, "An Attack Aircraft That's Cheap, Good, Gets Cold Shoulder," *The Wall Street Journal,* May 30, 1974, p. 1.
11. Roger Bilstein, *Flight in America,* rev. ed., Johns Hopkins Press, Baltimore, 1993, p. 292; and Ronald D. Green, *Brazilian Government Support for the Aerospace Industry,* U.S. Department of Commerce, March 1987, p. *x.*
12. David M. North, "Piper Gears for Strong Market Assault," *Aviation Week & Space Technology,* November 20, 1978, p. 97.
13. McDaniel, op. cit., p. 364; and *The Wall Street Journal,* October 4, 1973, p. 9.
14. McDaniel, op. cit., p. 366.
15. Ibid., p. 429.
16. G. Christian Hill and Barbara Isenberg, "Dangerous Planes: Testimony, Documents Indicate 4 Beech Models Had Unsafe Fuel Tanks," *The Wall Street Journal,* July 30, 1971, pp. 1 and 23.
17. McDaniel, op. cit., p. 465.
18. *Jane's All the World's Aircraft, 1976–1977,* p. 254.
19. Barry Bluestone, Peter Jordan, and Mark Sullivan, *Aircraft Industry Dynamics: An Analysis of Competition, Capital, and Labor,* Auburn House, Boston, 1981, p. 110.
20. Ethell, op. cit., p. 63.
21. "Crosswinds," p. 46.
22. McDaniel, op. cit., p. 425.
23. Warren J. Alverson, "The Shaky Case for the Company Jet," *Business Horizons,* Fall, 1972, p. 87.

Chapter 7

1. "500 Largest Industrial Corporations," *Fortune,* May 5, 1980, pp. 274–299.
2. *General Aviation Statistical Databook, 1990–1991,* p. 8.
3. Office of Aerospace, *A Competitive Assessment of the U.S. General Aviation Aircraft Industry,* Executive Summary, U.S. Department of Commerce, June 1986, p. *x*.
4. Erwin J. Bulban, "Raytheon, Beech Move Toward Merger," *Aviation Week & Space Technology,* October 8, 1979, p. 24.
5. McDaniel, op. cit., p. 509.
6. Ruth Simon, "Flying on a Wing and a Half," *Forbes,* July 13, 1987, p. 350; and "Why Beech Is Floating on Cloud 9," *Business Week,* March 19, 1990, p. 129.
7. Ian Goold, "Waxing Mooney," *Flight International,* November 29, 1986, pp. 105–106.
8. "British Aerospace Lawsuit Forces Fairchild to Seek Chapter 11 Protection," *Aviation Week & Space Technology,* February 19, 1990, p. 20.
9. "Potential Buyers Express Interest in Fairchild," *Aviation Week & Space Technology,* July 23, 1990, p. 34.
10. David M. North, "Gulfstream Terminates Production of Turboprop Commander Line," *Aviation Week & Space Technology,* January 28, 1985, p. 24.
11. "For Gulfstream, the Sky May Not Be the Only Limit," *Business Week,* February 26, 1990, p. 48.
12. *Interavia,* August 8, 1985, p. 845.
13. Ronald J. Wanttaja, *Kitplanes,* 2d ed., McGraw-Hill, New York, 1996, p. 7.
14. Ibid., pp. 22–24.
15. Author's interview with Danny Hiner, Vice President, Production, Aviat, November 21, 1997.
16. Experimental Aircraft Association statistical reports, from FAA figures.
17. Timothy K. Smith, "Liability Costs Drive Small-Plane Business Back into Pilots' Barns," *The Wall Street Journal,* December 11, 1991. p. A1.
18. Office of Aerospace, *A Competitive Assessment of the U.S. Civil Helicopter Industry,* U.S. Department of Commerce, April 1988, p. 68.
19. *Flying,* May 1991, p. 12.

20. McDaniel, op. cit., p. 383.
21. Author's interview with Stan Green, former vice president and general counsel of GAMA, May 27, 1992.
22. "Pulling Out of a Nose-Dive," *Economist,* June 18, 1988, p. 90.
23. Timothy K. Smith, op. cit., p. A1.
24. Office of Aerospace, *A Competitive Assessment of the U.S. General Aviation Aircraft Industry,* p. 53.
25. *Jane's All the World's Aircraft, 1989–1990,* p. 29.
26. "Pulling Out of a Nose-Dive," p. 90.
27. L. J. Truitt and S. E. Tarry, "The Rise and Fall of General Aviation: Product Liability, Market Structure, and Technological Innovation," *Transportation Journal,* Summer 1995, p. 61.
28. Office of Aerospace, *A Competitive Assessment of the U.S. General Aviation Aircraft Industry,* p. 22. The four firms involved were Beech, Cessna, Gulfstream, and Learjet.
29. Truitt and Tarry, op. cit., pp. 53, 61, and 65.
30. Timothy K. Smith, op. cit., p. A10.
31. "Against the Law," *Economist,* June 10, 1989, p. 66.
32. *General Aviation Statistical Databook, 1990–1991,* p. 4.
33. General Aviation Manufacturers Association, news release, January 16, 1992.
34. Office of Aerospace, *A Competitive Assessment of the U.S. General Aviation Aircraft Industry,* p. 10.

Chapter 8

1. *Jane's All the World's Aircraft, 1990–1991,* p. 475.
2. Truitt and Tarry, op cit., p. 61.
3. *Jane's,* 1990–1991, p. 393.
4. "Against the Law," p. 66.
5. "Why Beech Is Floating on Cloud 9," pp. 128–129.
6. "Piper May Still Be Carrying Excess Baggage," *Business Week,* June 12, 1989, p. 76.
7. "Pulling Out of a Nose-Dive," p. 91.
8. "Piper May Still Be Carrying Excess Baggage," p. 76.
9. "Without Cash, Piper May Have Trouble Keeping Its Nose Up," *Business Week,* March 5, 1990, p. 32; and Edward A. Phillips, "Piper Lays Off 170 Employees, Reduces Aircraft Production," *Aviation Week & Space Technology,* July 23, 1990, p. 32.

10. James T. McKenna, "Piper Creditors to Vote on Reorganization Plan," *Aviation Week & Space Technology,* May 22, 1995, p. 24.
11. Scott Thurston, "FedEx Deal Could Put Small Firm 'on the map,'" *Atlanta Journal/Constitution,* December 1, 1996, p. R6.
12. Eric Weiner, "For Now, the Ultimate Status Symbol Still Sells," *The New York Times,* September 30, 1990, p. C4.
13. G. Pierre Goad, "Salvaged Units Help Bombardier Sell Jets," *The Wall Street Journal,* April 26, 1990, p. B11.
14. *Flying,* June 1991, p. 110.
15. John R. Wilke, "Beech's Sleekly Styled Starship Fails to Take Off with Corporate Customers," *The Wall Street Journal,* September 29, 1993, p. B1.
16. Paul Proctor, "Raytheon Restructures Merged Aircraft Units," *Aviation Week & Space Technology,* June 5, 1995, p. 58.
17. Scott Thurston, "Overhauled Gulfstream Preparing to Go Public," *Atlanta Journal/Constitution,* September 3, 1996, P. C4.
18. *General Aviation Statistical Databook, 1990–1991,* p. 24.
19. Ibid., p. 11.
20. *General Aviation Statistical Databook, 1996,* GAMA, Washington, D.C., p. 11.
21. "Why Beech Is Floating on Cloud 9," pp. 128–129.
22. Rod Simpson, "From Beagles to Beechjets: General Aviation Defined," *Air International,* October 1995, p. 207.
23. Paul Proctor, "Helicopter Sales Buoyed by Strong Economics," *Aviation Week & Space Technology,* February 3, 1997, pp. 54–56.
24. Edward H. Phillips, "Advanced Kit-Builts Dominate EAA Show," *Aviation Week & Space Technology,* August 15, 1994, p. 50.
25. *Annual Industry Review,* GAMA, Washington, D.C., 1997, p. 1.
26. Ibid., p. 5.

Bibliography

Articles, Monographs, and Reports

"Cessna Airplane Factory Is Located at Wichita," *Aviation and Aeronautical Engineering,* December 1, 1916, p. 294.

"Fairchild Company Buys Control of Kreider-Reisner," *Wings of Industry,* April 15, 1929, p. 1.

"Count the Cubs," *Fortune,* June 1940, pp. 77–81; 114–118.

"New Planes for Personal Flying," *Fortune,* February 1946, pp. 124–129; 154–158.

"Experiment at Republic," *Fortune,* February 1946, pp. 123–125; 164–172.

"Beechcraft Takes Off on the Wings of a Jet," *Business Week,* February 25, 1956, pp. 178–186.

"Aero Commander Plans to Market Full Single-Engine Aircraft Line," *Aviation Week & Space Technology,* March 2, 1964, p. 36.

"A Divided House at Piper," *Business Week,* March 16, 1974, pp. 94–97.

"Aviation Pioneers," *Wichita Eagle-Beacon,* 1987 reprint of a series of articles on Kansas aviation history published in the *Eagle-Beacon,* beginning October 8, 1984.

"Pulling Out of a Nose-Dive," *Economist,* June 18, 1988, pp. 90–91.

"Against the Law," *Economist,* June 10, 1989, p. 56.

"Piper May Still Be Carrying Excess Baggage," *Business Week,* June 12, 1989, p. 76.

"British Aerospace Lawsuit Forces Fairchild to Seek Chapter 11 Protection," *Aviation Week & Space Technology,* February 19, 1990, p. 20.

"For Gulfstream, the Sky May Not Be the Only Limit," *Business Week,* February 26, 1990, p. 48.

"Without Cash, Piper May Have Trouble Keeping Its Nose Up," *Business Week,* March 5, 1990, p. 32.

"Why Beech Is Floating on Cloud 9," *Business Week,* March 19, 1990, pp. 128–129.

"Potential Buyers Express Interest in Fairchild," *Aviation Week & Space Technology,* July 23, 1990, p. 34.

Aerospace, Office of: *A Competitive Assessment of the U.S. Civil Helicopter Industry,* Department of Commerce, April 1988.

Alverson, Warren J.: "The Shaky Case for the Company Jet," *Business Horizons,* vol. 15, April 1972, pp. 79–88.

Bianco, Anthony: "Gulfstream's Pilot," *Business Week,* April 14, 1997, pp. 64–76.

Brown, David A.: "Aero Commander Management Shifts Set," *Aviation Week & Space Technology,* July 19, 1965, pp. 20–21.

Burner, David L.: "GAMA Agenda: Address to General Aviation Manufacturers Association," February 10, 1995.

Crouch, Tom.: "General Aviation: The Search for a Market," in Eugene M. Emme (ed.), *200 Years of Flight in America: A Bicentennial Survey,* Univelt, San Diego, 1977, pp. 111–135.

Department of Commerce: *A Competitive Assessment of the U.S. General Aviation Aircraft Industry,* June 1986.

Foster, John, Jr.: "The Personal Plane Sales Target," *Aviation,* January 1994, pp. 116–117 and 364–368.

Geisse, John H., and Samuel C. Williams: *Postwar Outlook for Private Flying,* Report to Assistant Secretary of Commerce W. A. M. Burden, 1943.

Goad, G. Pierre: "Salvaged Units Help Bombardier Sell Jets," *The Wall Street Journal,* April 26, 1990, p. B11.

Goold, Ian: "Waxing Mooney," *Flight International,* November 29, 1986, pp. 105–106.

Green, Ronald D.: *Brazilian Government Support for the Aerospace Industry,* Department of Commerce, March 1987.

Hall, Randolph F.: "Cunningham-Hall Aircraft Corporation Story," *American Aviation Historical Society Journal,* Summer 1971, pp. 90–97.

Hardie, George, Jr.: "The Long Road Back," *Sport Aviation,* June 1986, pp. 35–37.

Harris, Thomas J.: *The Magic of Aero Design,* The Newcomen Society, New York, May 10, 1962 (Monograph).

Hedrick, Frank E.: *Pageantry of Flight: The Story of Beech Aircraft Corporation,* The Newcomen Society, New York, September 28, 1957 (Monograph).

Hill, G. Christian, and Barbara Isenberg: "Dangerous Planes: Testimony, Documents Indicate 4 Beech Models Had Unsafe Fuel Tanks," *The Wall Street Journal,* July 30, 1971, pp. 1 and 23.

Klemin, Alexander: "Investing in Aviation," *Scientific American,* February 1929, pp. 148–153.

Levine, Richard J.: "An Attack Aircraft That's Cheap, Good, Gets Cold Shoulder," *The Wall Street Journal,* May 30, 1974, p. 1.

McKenna, James T.: "Piper Creditors to Vote on Reorganization Plan," *Aviation Week & Space Technology,* May 22, 1995, p. 24.

Matt, Paul R.: *Aeronca: Its Formation and First Aircraft,* vol. 10: *Historical Aviation Album,* 1971, pp. 270–285.

———: *Alfred Victor Verville,* vol. 18: *Historical Aviation Album,* 1987, pp. 88–96.

Mayborn, Mitch, and Peter M. Bowers: "A History of the Stearman Aircraft Company," *Stearman Guidebook,* Flying Enterprise Publications, Dallas, 1973, pp. 4–5.

Nevill, John: "The Story of Wichita," *Aviation,* September 1930, pp. 166–170; November 1930, pp. 291–295; and December 1930, pp. 353–357.

North, David M.: "Piper Gears for Strong Market Assault," *Aviation Week & Space Technology,* November 20, 1978, p. 97.

Phillips, Edward H.: "Piper Lays Off 170 Employees, Reduces Aircraft Production," *Aviation Week & Space Technology,* July 23, 1990, p. 32.

———: "Advanced Kit-Builts Dominate EAA Show," *Aviation Week and Space Technology,* August 15, 1994, p. 50–51.

Piper, W. T., Jr.: *From Cubs to Navajo: The Story of Piper Aircraft Corporation,* The Newcomen Society, New York, April 23, 1970 (Monograph).

Proctor, Paul.: "Raytheon Restructures Merged Aircraft Units," *Aviation Week & Space Technology,* June 5, 1995, pp. 58–59.

———: "Helicopter Sales Buoyed by Strong Economics," *Aviation Week & Space Technology,* February 3, 1997, pp. 54–56.

Rankin, Robert H.: "The Genius of Giuseppe," *Flying,* February 1953, pp. 36–37 and 60–61.

Roos, Frederick W.: "Curtiss-Wright St. Louis," *American Aviation Historical Society Journal,* vol. 35, Winter 1990, pp. 293–305.

Schneider, Charles E.: "Piper Faces Fiscal, Internal Hurdles," *Aviation Week & Space Technology,* October 19, 1970, pp. 67–69.

Schreiner, Herm: "The Waco Story, Part 1: Clayton Brukner & the Founding Years," *American Aviation Historical Society Journal,* vol. 25, Winter 1980, pp. 281–299.

———: "The Waco Story, Part 2: Expansion with the 'F' Series," *American Aviation Historical Society Journal,* vol. 29, Fall 1984, pp. 214–227.

Simon, Ruth: "Flying on a Wing and a Half," *Forbes,* July 13, 1987, pp. 350 and 355.

Simpson, Rod: "From Beagles to Beechjets: General Aviation Defined," *Air International,* October 1995, pp. 205–209.

Smith, Timothy K.: "Liability Costs Drive Small-Plane Business Back into Pilots' Barns," *The Wall Street Journal,* December 11, 1991, pp. A1 and A10.

Solow, Herbert: "North American: A Corporation Deeply Committed," *Fortune,* June 1962, pp. 145–149 and 164–182.

Stevenson, David B.: "Charles Healy Day and His New Standards," *American Aviation Historical Society Journal,* vol. 41, Fall 1996, pp. 200–219.

Thurston, Scott: "Overhauled Gulfstream Preparing to Go Public," *Atlanta Constitution,* September 3, 1996, p. C4.

———: "FedEx Deal Could Put Small Firm 'On the Map,'" *Atlanta Journal/Constitution,* December 1, 1996, p. R6.

Townsend, Lew: "Lightplane Mergers: Beech, Cessna, Lear Didn't Get Together," *Wichita Eagle-Beacon,* February 10, 1990, pp. 1F and 8F.

Trimble, William F.: "The Collapse of a Dream: Lightplane Ownership and General Aviation in the United States after World War II," in William F. Trimble (ed.) *Pioneers and Operations,* vol. 2: *From Airships to Airbus: The History of Civil and Commercial Aviation,* Smithsonian Institution Press, Washington, D.C., 1995, pp. 128–145.

Truitt, L. J., and S. E. Tarry: "The Rise and Fall of General Aviation: Product Liability, Market Structure, and Technological Innovation," *Transportation Journal,* Summer 1995, pp. 52–70.

Vidal, Eugene L.: "Low-Priced Airplane," *Aviation,* February 1934, pp. 40–41.

Weiner, Eric.: "For Now, the Ultimate Status Symbol Still Sells," *The New York Times,* September 30, 1990, p. F4.

West, Ted: "Why Wichita," *Flying,* September 1997, p. 254.

Whittier, Bob: "Light Plane Heritage," *EAA Experimenter,* December 1993, pp. 29–34.

Wilke, John R.: "Beech's Sleekly Styled Starship Fails to Take Off with Corporate Customers," *The Wall Street Journal,* September 29, 1993, pp. B1 and B8.

Wilkinson, Kim M.: "History of the Piper Aircraft Corporation," *Lock Haven Express,* July 14–18, 1987.

Williams, Nicholas M.: "The Aero Commander 520," *American Aviation Historical Society Journal,* vol. 35, Spring 1990, pp. 18–37.

Wilson, Kenneth D., and Thomas E. Lowe: "Lloyd C. Stearman, 1898–1975," *American Aviation Historical Society Journal,* vol. 36, Summer 1991, pp. 82–93.

Books

Austin, Edward T.: *Rohr: The Story of a Corporation,* Rohr Corporation, Chula Vista, CA, 1969.

Bilstein, Roger: *Flight Patterns,* University of Georgia Press, Athens, GA, 1983.

———: *Flight in America,* Johns Hopkins University Press, Baltimore, 1984, revised edition 1993.

———, and Jon Miller: *Aviation in Texas,* Texas Monthly Press, Austin, 1985.

Cessna Aircraft Company: *An Eye to the Sky,* Cessna Aircraft Company, Wichita, 1962.

Corn, Joseph J.: *The Winged Gospel: America's Romance with Aviation,* Oxford University Press, New York, 1983.

Cunningham, William G.: *The Aircraft Industry: A Study of Industrial Location,* Lorrin L. Morrison, Los Angeles, 1951.

Ethell, Jeffrey L.: *NASA and General Aviation,* NASA, Washington, D.C., 1986.

Fairchild Hiller Corporation: *Yesterday, Today and Tomorrow: Fifty Years of Fairchild Aviation,* Fairchild Hiller Corporation, Germantown, Md., 1970.

Francis, Devon: *Mr. Piper and His Cubs,* Iowa State University Press, Ames, IA, 1973.

Gunston, Bill: *One of a Kind: The Story of Grumman,* Grumman Corporation, Bethpage, NY, 1988.

Hallion, Richard P.: *Designers and Test Pilots,* the *Epic of Flight* series, Time, Inc., New York, 1983.

Harding, William Barclay: *The Aviation Industry,* Charles D. Barney, New York, 1937.

Horgan, James J.: *City of Flight: The History of Aviation in St. Louis,* The Patrice Press, Gerald, MO, 1984.

Lambermont, Paul, with Anthony Pirie: *Helicopters and Autogiros of the World,* rev. ed., A. S. Barnes and Co., New York, 1970.

Loening, Grover C.: *Takeoff into Greatness: How American Aviation Grew So Big So Fast,* Putnam, New York, 1968.

Marcovski, Michael A.: *ARV—The Encyclopedia of Aircraft Recreational Vehicles,* Aviation Publishers, Hummelstown, PA, 1984.

McDaniel, William H.: *The History of Beech,* Beech Aircraft Corporation, Wichita, 1982.

National Advisory Committee for Aeronautics: *NACA-Industry Conference on Personal Aircraft Research,* NACA Langley Memorial Aeronautical Laboratories, Langley Field, VA, September 20, 1946.

Peek, Chet: *Taylorcraft: The Taylorcraft Story,* SunShine House, Terre Haute, IN, 1992.

Petersen, Carl J.: *The CallAir Affair: An Aeronautical History,* (privately published by the author), 1989.

Phillips, Edward H.: *Travel Air: Wings over the Prairie,* Flying Books, Eagan, MN, 1982.

———: *Cessna: A Master's Expression,* Flying Books, Eagan, MN, 1985.

———: *Beechcraft: Pursuit of Perfection,* Flying Books, Eagan, MN, 1992.

Rasche, Richard: *Stormy Genius,* Houghton Mifflin, Boston, 1985.

Rowe, Frank Joseph, and Craig Miner: *Borne on the South Wind: A History of Kansas Aviation,* The Wichita Eagle and Beacon Publishing Co., Wichita, 1994.

Smith, Frank Kingston: *Legacy of Wings: The Story of Harold F. Pitcairn,* J. Aronson, New York, 1981.

Swick, John C.: *The Luscombe Story: Every Cloud Has a Silvaire Lining,* SunShine House, Terre Haute, IN, 1987.

Thruelson, Richard: *The Grumman Story,* Praeger, New York, 1976.

Trimble, William F.: *High Frontier: A History of Aeronautics in Pennsylvania,* University of Pittsburgh Press, Pittsburgh, 1982.

Underwood, John W.: *The Stinsons,* The Heritage Press, Glendale, CA, 1969.

Wagner, William: *Ryan, the Aviator,* McGraw-Hill Book Co., New York, 1971.

Wanttaja, Ronald J.: *Kitplane Construction,* 2d ed., McGraw-Hill, New York, 1996.

Woods, George Bryant: *The Aircraft Manufacturing Industry: Present and Future Prospects,* White, Weld, New York, 1946.

Annuals and Directories

World Aviation Annual 1948, Aviation Research Institute, Washington, D.C., 1948.

Aerospace Industries Association: *Aviation Facts and Figures* and *Aerospace Facts and Figures,* Aerospace Industries Association, Washington, D.C., (various editions).

Aerospace Industries Association: *Aircraft Year Book* and *Aerospace Year Book,* Aerospace Industries Association, Washington, D.C., (various editions).

General Aviation Manufacturers Association: *General Aviation Statistical Databook,* GAMA, Washington, D.C., (various editions).

———: *Annual Industry Review: 1997 Outlook and Agenda,* GAMA, Washington, D. C., 1997.

Jane's All the World's Aircraft, Sampson Low, Marston and Co., London, (various editions).

Documentary Collections

Walter H. and Olive Ann Beech Papers, Ablah Library, Wichita State University, Wichita, Kansas.

Clayton J. Brukner Collection, Wright State University Archives and Special Collections, Wright State University, Dayton, Ohio.

International Cyclopedia of Aviation Biography, Wright State University Archives and Special Collections, Wright State University, Dayton, Ohio.

Index

NOTE: Listings are not strictly in alphabetical order. For clarity, names of company founders are placed first, followed by the company name if incorporating that of the founder, then by the aircraft of that manufacturer. Similarly, aircraft designations or model numbers of a particular manufacturer may not be strictly numerical but placed in approximate order of appearance. With the numerous mergers/acquisitions and transfers of design/production rights within the industry, some judgments have been made in placing particular aircraft under particular companies, and in separate listings for variants of basic designs. In certain instances an aircraft type is duplicated under different manufacturers for completeness. Information in tables, illustrations, and in the chronology is not indexed.

A

D

G

H

N

About the Author

Donald M. Pattillo (Acworth, Georgia) is an expert in both international business and the aerospace industry. He taught domestic and international business administration at Northeastern University, the University of Dayton, St. John's University, and Point Park College. He has written articles on the aerospace industry in world trade and aerospace industry financial management. Dr. Pattillo's first book, *Pushing the Envelope* (University of Michigan Press, Spring 1998), is a business history of the military/commercial aerospace industry.